FORMULATING AND PROCESSING DIETETIC FOODS

First Edition

Other books edited or written by the same author —

Chemistry and Technology of Cereals as Food and Feed
First and Second Editions

Technology of New Product Development

Glossary of Milling and Baking Terms

Bakery Technology and Engineering
First, Second, and Third Editions

Equipment for Bakers

Cereal Technology

Cereal Science

Food Texture

Water in Foods

Bakery Technology

Ingredients for Bakers
First and Second Editions

Cookie and Cracker Technology
First, Second, and Third Editions

Technology of the Materials of Baking

Glossary of Cereal Science and Technology

Snack Food Technology
First, Second and Third Editions; Japanese Edition

* * * * * *

FORMULATING AND PROCESSING DIETETIC FOODS

by

SAMUEL A. MATZ, PH. D.

President, Pan-Tech International, Inc. Formerly, Vice President for Research, Development, and Compliance, Ovaltine Products, Inc. At one time, Vice President for Research and Development, Robert A. Johnston Co.; Technical Director of the Refrigerated Dough Program, Borden Foods Co.; Chief of the Cereal and General Products Branch, Quartermaster Food and Container Institute for the Armed Forces; Chief Chemist, Harvest Queen Mill and Elevator Co.; Instructor, Department of Flour and Feed Milling Industries, Kansas State University; Chemist, Iglehart Mills Division of General Foods.

PAN-TECH INTERNATIONAL, INC.
P. O. BOX 4548
MC ALLEN, TEXAS 78502
1996
Phone & Fax 956-702-7085

Formulating and Processing Dietetic Foods

Copyright 1996 by SAMUEL A. MATZ

ISBN 0-942849-28-0

Pan-Tech International, Inc.
P. O. Box 4548
McAllen, TX 78502

Printed in the United States of America

PREFACE

This book was written to provide food technologists and scientists with information on the actual and perceived need of consumers for special dietary foods and to give the basic guidance necessary for fulfilling these requirements with new or existing products.

About a hundred prototype formulas and nearly the same number of processes are given for dietetic foods. Some of these are of conventional types while others are experimental, and still others require ingredients that are currently not approved for food use but which may become acceptable in the near future. The latter should be regarded as more or less speculative.

In-depth discussions of the medical conditions and symptoms resulting from, or said to result from, nutritional deviations from the norm are given for obesity, high blood pressure, diabetes, atherosclerosis, and other diseases. The author takes a skeptical view of much of the conventional wisdom in the dietary management realm, but acknowledges the marketing power of many of the concepts presented to us daily by the news and entertainment media.

Among the many types of dietetic food covered in these pages are low-calorie, low sugar, low sodium/salt, non-allergenic, reduced/no cholesterol, and high fiber varieties. Conditions that these foods are expected to alleviate or prevent include obesity, high blood pressure, atheroclerosis, coronary heat disease, asthma, diabetes, neural tube defects. and cancer. In addition, sports bars and drinks, metered foods, and non-cariogenic ingredients are discussed.

Philosophical and religious considerations affecting the dietary choices of signicant numbers of Americans are also covered. Supplements, including herbal types of material that are alleged to have special curative or prophylactic properties are described, as well as vitamin, mineral, protein, and fiber fortifiers.

Although the book is intended mainly for persons actively engaged in developing or producing dietetic products, it should also have considerable value for a wider audience. It should be useful to food technologists who are connected with industries that supply raw materials for manufacturers of these products. And, marketing and sales personnel may find the information useful in providing an insight into the needs and expectations of customers, and possibly in revealing future trends.

At the end of the book, the reader will find a glossary of terms that are either unfamiliar to most food technologists or have special meanings within the context of this volume. There is also a bibliography with about 400 entries and an eight-page index.

The bibliography will be particularly useful for those readers who wish to make an in-depth exploration of all possibilities connected with a highly specialized problem or opportunity. Most of the references are from articles published since 1990, and about 10% are from 1996 publications.

Government and state regulations are always important and may, at times, be overriding considerations in designing, labeling, and marketing dietetic foods. Much space has been devoted to discussions of these obstacles, but conclusive answers to individual problems cannot always be given. Because of the constantly changing nature of the legal landscape, and its extreme complexity, the innovator is advised to consult current government publications, which are available from government bookstores or the Office of the Superintendent of Documents. In addition, many libraries either stock the major references or can obtain them as interlibrary loans.

The author welcomes comments from readers.

Samuel A. Matz
August 31, 1996
Alamo, Texas

TABLE OF CONTENTS

chapter one

INTRODUCTION

Foods can be divided into two major categories: (1) those selected by the consumer because they provide some combination of organoleptic satisfaction, economic advantage, or general nutrition that is attractive at the time of purchase, and (2) dietetic foods — those selected by the purchaser because they meet some special nutritional need — a need that is not primarily influenced by economic or esthetic considerations or the immediate satisfaction of hunger. It is the second category that will be covered in this book.

CATEGORIES OF DIETETIC FOODS

For the purposes of the present discussion, it is useful to regard dietetic foods as being divided into two major categories:

(1) Foods intended to confer a health benefit by modifying some special facet of the body's physiological processes or by avoiding unwanted interference with those processes. It is characteristic of foods in this category that they are subject to legal requirements governing their manufacture, distribution, label claims, and other aspects of their production, sale, and use. There are at least three types of foods in this category, those which:

(a) Eliminate components that cause physical harm to persons having certain functional inadequacies (e.g., diabetes) or hypersensitivities (e.g., allergies).

(b) Enable essentially normal persons to enhance a somatic characteristic, as by increasing muscular development or reducing body weight.

(c) Meet specialized needs of persons affected by some unusually stressful but temporary condition: pregnancy, trauma (accident or surgery), severe or prolonged illness, etc.

(2) Foods that conform to religious or philosophical tenets. Although persons adhering to such spiritual guidelines may claim enhanced physical power, better mental or emotional status, or prolonged life as a result of their practices, the dietary principles they follow are characterized by a lack of objective evidence that the diet has a beneficial effect on the physiological status of the practitioner. These dietary regimens are generally based on revelations received directly or indirectly from a divine being or a human authority claiming special and unique extrasensory knowledge, whether or not they are publicly characterized as such. Such foods may be classified as:

(a) Foods consumed or avoided as a result of a divinely revealed principle and traditional foods that have a religious component. The most widely recognized example of this category, at least in the US, is kosher food. Also in this group are foods conforming to Muslim, Adventist, Hindu tenets, etc.

(b) Foods consumed or avoided as a result of philosophical beliefs that have not yet reached the level of a structured set of divine commands. A common example is vegetarianism, of which there are several forms.

Foods having as their principal appeal that of conforming in a general way to ethnic traditions of certain groups of consumers, e.g., Irish soda bread, bagels, and tortillas, but which do not have an organized philosophical structure behind the tradition, are discussed under other categories, or not discussed at all.

COMMERCIAL TYPES OF DIETETIC FOODS

The designer of dietetic foods who is working in a commercial environment generally has to deal only with products of the following four types, or sub-categories:

(1) Foods that facilitate weight reduction or maintenance of desired weight in persons who do not have detectable physiological impediments, such as:

(a) Low-calorie foods that can be substituted for conventional or ordinary foods which are being consumed in excessive amounts.

(b) Metered foods of conventional formulation that make it easier to control intake within a targeted range of calories, the food often containing a given percentage of all, or most, essential nutrients in a single container so that the consumer can obtain all his nutritional needs with a minimum amount of calorie intake.

(c) Foods specially engineered to promote satiety without excess calories, such as high fiber foods.

(2) Foods that will delay or prevent the development of diseased conditions other than obesity, such as atherosclerosis.

(3) Foods that will not cause adverse reactions in persons with aberrant physiological status, such as:

(a) Foods from which specified allergens have been eliminated.

(b) Foods containing insignificant amounts of (or no) ingredients to which some persons are sensitive but not allergic.

(4) Foods that will meet special and (usually) temporary physiological needs due to the influence of unusually stressful conditions.

(a) Sports foods for increasing endurance, building muscle, etc.

(b) Foods for invalids, victims of surgery, pregnant and lactating women, and other other persons with special nutritional needs due to a temporary physiological status.

(5) Foods and beverages designed to conform to religious tenets or philosophical principles.

(a) Foods meeting the rules imposed by authorities of religions having dietary restrictions, such as Jewish, Muslim, and Seventh Day Adventist.

(b) Foods meeting the requirements of one of the sets of vegetarian guidelines, etc.

This book deals to some extent with all of these kinds of foods, but considerable emphasis is placed on weight-control products. This policy has been based on the recognition that, both as a health problem and as a commercial opportunity, foods and beverages that help reduce the incidence of obesity are more important to more people than any other category of dietetic foods.

THE PROBLEM OF OBESITY

Of all the health problems confronting the individual in today's environment, the most common one is overweight. A large proportion of the population is faced with this problem for most of their lives. And, it is a serious problem, causing numerous health disorders as well as reducing the quality of life. Many persons recognize that they are overweight, or have a tendency to gain unwanted pounds, and they want help. Assisting them in controlling their weight is the task of this treatise. Many of the chapters in this book deal with foods that are intended to help persons control their weight.

Controlling Body Weight

For a healthy individual — one whose physiological status is within normal ranges — excess body weight can be reduced by increased exercise, decreased consumption of food, or consuming approximately the same volume and weight of food to which he has become accustomed but choosing foods from those known to be of lower calorie content than the customary versions. Increased exercise and reduced food consumption are generally considered the least desirable of the three options. Substituting lower calorie versions of the foods presently being consumed is generally considered the least objectionable alternative. Consequently, there is a large demand for products that resemble well-liked and commonly eaten foods but contain fewer calories per portion.

Cutting back on the actual volume of food ingested is more difficult and troubling than using the same size portions (and same number of portions) but selecting foods that have been designed to provide fewer calories in the same size portion that the consumer is accustomed to consume. Weight-control dieters often make the mistake of changing only a part of their diet to low-calorie foods, leaving intact their practice of consuming too much total foods. It has often been pointed out that one's entire food intake, not a single part of it, is the critical variable determining nutritional adequacy of a diet. It is also the whole diet, not a part of it, that satisfies the psychological needs of the individual.

Health Problems Connected With Obesity

The prevalence of obesity — From the Surgeon General's Report, we learn that approximately 25% of adult women and 42% of adult men are considered overweight, 14% of women and 12% of men are considered obese. Many children are also seriously overweight. Estimates from other sources differ from this report, perhaps because there are different criteria in use for judging obesity. The weight in pounds as related to height in inches is a familiar method of reporting, and the charts carrying these figures usually have subdivisions based on age groups, gender, and, sometimes, body build.

The body mass index (BMI) is being more frequently used in reports of weight problems. BMI consists of the ratio of total weight (in kilograms) to the square of the height (in square meters). It is no part of the present discussion to dwell on the consistency or rationality of such a measure as an indicator of proper physiological status relative to body composition.

Persons with a BMI in the 20 to 25 kg/m^2 range are considered normal in weight, those in the 25 to 30 category are "overweight," if in the 30 to 40 range are considered "obese," while those with a BMI of more than 40 are regarded as being "severely obese." The categories change with age: a "normal" BMI for adults aged 19 to 24 years has been defined as 19 to 24, but this increases to 24 to 29 for adults over age 65. Obesity has also been defined as a body fat content greater than 30% in women and 20% to 25% in men — by this standard, almost all men and women over the age of 50 would be considered obese.

Effects of obesity — In addition to its depressing effect on the general quality of life, obesity increases the risk of contracting many diseases. It shows a positive correlation with the frequency of diabetes, hyperlipidemia, hypercholesterolemia, gallstones, cancer, and hypertension. Many other diseases seem to be more common in overweight persons. Women who are obese around the time of conception run a much greater risk of having babies that have defects of the spinal cord. In 1988, the US Surgeon General issued a report entitled "Nutrition and Health" which summarized the scientific evidence for the role of diet in health and disease. It concluded, among other things, that excessive consumption of foods, especially certain types of foods, should be a concern of Americans.

A Scientific Status Summary published by the Institute of Food Technologists' Expert Panel on Food Safety and Nutrition (edited by LaChance 1994) listed the following major health and social problems thought to be caused or exacerbated by obesity:

Coronary heat disease
Stroke
Obstructive sleep apnea
Diabetes mellitus
Gout
Hyperlipidemia
Osteoarthritis
Reduced fertility
Impaired psychosocial function
Reduced agility and increased risk of accidents
Impaire obstetrical performance
Reduced economic performance
Encountering discrimination and prejudice

Factors causing obesity — It is quite clear that several demographic factors are predictive of the ratio of obese persons. In the US, obesity is more common in the Northeast and Midwest than elsewhere. It is higher in metropolitan areas than in sparsely populated regions. Ethnic affiliation often shows a connection. In the US, the proportions of women whose weight is 20% or more above the desirable level are 35% for African-Americans, 26% for Hispanic, and 20% for non-Hispanic white women. Gender is an important factor, with women being obese at about twice the rate for men.

Table 1.1
RISK FACTORS FOR ADULT OBESITY

Demographic factors	Personal factors
Country of residency	Obesity in family
Region of country	Childhood obesity
Urban or rural	Cigarette smoking
Racial group	Drugs
Gender	Childbearing
Socioeconomic status	Menopause
Education level	Physical inactivity
Age	Diet high in fat

Source: modified from LaChance (1994)

Obesity prevalence increases with age of the population. Low economic status is also associated with a higher incidence of obesity. If an individual was obese in childhood, he has a greater risk of obesity than those who were of average weight. One author found that 33% of adults found to be obese at 36 years of age had been overweight at age 7, and 63% had been obese by age 14. Childbearing and menopause are also risk factors; typically, not all the fat gained during a pregnancy is lost postpartum.

Animal tests have provided overwhelming evidence that a tendency to obesity can be transmitted genetically. The fat gain in genetically obese rats and mice is related to increased consumption of food and less expenditure of energy per unit of body weight. Other factors are also involved. At least some of these animals have defective hormone receptors in their liver and muscles. Animal tests and studies with humans have not provided unequivocal answers to all the questions involved in the genesis and progress of overweight.

GENERAL APPROACHES TO CALORIE REDUCTION

As we all know, cutting back on the actual volume (number of portions and/or sizes of portions) of food eaten during a given period is a difficult and unpleasant change, involving the breaking of long-established habits and withstanding the urges of hunger. It would be preferable to be able to consume the same size portions and the same number of portions while reducing caloric intake. Consequently, a very large business has grown up around the production of foods of low caloric density. The Calorie Control Council recently reported (Anon. 1996A) that more than 170 million Americans consume low-calorie and reduced-fat foods.

Fats and carbohydrates are usually the two types of substances consumed in excess by persons who are obese. Excess consumption of protein does not seem to be a major cause of overweight, though of course this dietary component does make its own significant contribution to the total caloric intake. Alcohol (at about 7 calories per gram) can also contribute to excess body weight — and alcoholic beverages are significant parts of the diet for many people — but it will not be discussed in this book.

Different food types present different problems when designing low-calorie products. Seldom is a solution found that is applicable to a wide range of products. Most non-bakery products that are advertised as having reduced calorie content depend upon the replacement of most of the fat (about nine calories per gram) either by water (no calories) or by carbohydrates and proteins (about four calories per gram). For example, in low calorie beverages and jellied puddings, the sugar or other caloric sweetener found in the conventional product has been replaced by water plus a very small amount of a potent artificial sweetener. If the normal product contains very little fat, significant calorie reduction per unit weight can only be achieved by replacing carbohydrates with water or with some essentially undigestible solid material such as fiber.

Because of the complex nature of bakery products and their sensitivity to variations in composition, the usual approaches to formulating reduced calorie versions are seldom practical. Significant reduction of calories per unit weight of white bread is very difficult because there is no non-caloric replacement ingredient for gluten or starch that does not have severe effects on the organoleptic characteristics of the product when it is included at the level required to make up the mass and volume expected in a traditional product. Merely replacing fats or carbohydrates with water is rarely, if ever, possible. The starch that forms such a large proportion of flour and bread cannot be replaced by any substance of significantly lower caloric content (less than, say, one calorie per gram) without destroying most of the desirable characteristics of bread.

Use of non-digestible substances such as purified cellulose (which theoretically contributes no calories to the consumer) to replace flour has received some attention. Other obvious candidates as replacements are ingredients having high dietary fiber contents such as corn bran or rice bran. Unfortunately, these materials also have pronounced effects on the physical and sensory properties of the dough and bread, so much so that even the most dedicated weight-watcher might find the food unacceptable. Also, white bread is a product controlled by Federal Standards of Identity and virtually nothing of a calorie-reducing nature can be accomplished without violating these regulations. Designing the label so as to clearly indicate the product is not "white bread" is an approach that has been used to avoid the restrictions of the Federal Standards.

Another approach that has been tried is to reduce the calories per serving, not the calories per unit weight. The serving is defined either as a certain production-sized piece (one roll), or two slices, or the amount that has been packed in a single-serving container (such as a plastic pouch). The manufacturer then advertises a comparison between the calories per serving of the dietetic bread and the calories per serving weight (usually two slices) of regular white bread. Calorie reduction is accomplished by reducing the weight of a slice, usually by making it thinner but occasionally by modifying the height or width.

Although at first thought this seems to be the most brazen type of equivocation, there is actually some logic to the approach. For example: a large percentage of sliced bread is consumed as sandwich components and, presumably, only two slices per sandwich would be consumed regardless of whether a slice weighed one ounce or three-quarters of an ounce. Some manufacturers have gone a step further and attempted to puff up the bread so that a lighter weight slice has about the same dimensions as a slice of regular bread but weighs significantly less. There are regulatory uncertainties involved in all of these alternatives.

LABELING REGULATIONS AND OTHER LEGAL RESTRICTIONS

Federal, state, and local governments are partners, and not silent partners, in the development, production, and marketing of foods of all kinds, but their interference with dietetic foods is especially great. There are labeling restrictions affecting all of the categories that will be discussed, and the manufacturer must be prepared to meet government regulations if the intention is to offer such products for public sale (as opposed to, for example, prescription sale where different, but often more restrictive, regulations may or may not apply).

The National Labeling and Education Act

According to FDA regulations interpreting the NLEA of 1990, a finished food must contain no more than 40 calories per serving if a claim of "low-calorie" is to be made on the package label. If a product is claimed to be "low-fat" it must contain no more than three grams of fat per serving, while "fat-free" must be justified by the presence of no more than 0.5 gram of fat in a serving. Furthermore, a "reduced-calorie" product must contain at least 25% fewer calories than the standard, reference, or regular product. "Light" or "lite" refers to a reduction of 33% in calories or 50% fat when compared to the common, ordinary, usual, or reference type of food.

If a nutrient content claim is to be made, The Nutrition Labeling Education Act requires that the nutrient in question must be one that has been assigned a Reference Daily Intake (RDI), and the food must contain at least 20% of the RDI of the nutrient. Then, the manufacturer may claim the food is "high," "rich," or "an excellent source" of the nutrient. These restrictions are different from those that must be met when the labeling is to include health claims.

A Caution to the Reader

This book includes many examples of formulas for foods containing experimental ingredients (some of them taken from patents or other literature), that are intended as illustrations of possible future applications. These ingredients may not have been approved by regulatory agencies and cannot reasonably be considered as falling into the category of foods Generally Recognized as Safe (GRAS). Therefore, their commercial use remains somewhere in the indefinite future.

DATABASES

Printed Databases

There are a number of paper and electronic databases useful to the designer of dietetic foods. The major semi-official compendium of data on the nutrient content of foods (and some ingredients) is USDA's Handbook *8*, now amounting to about twenty loose-leaf volumes, intended to ultimately cover nearly the whole universe of edibles available to the American consumer. Individual volumes can be obtained from the Superintendent of Documents, Washington DC. If there is a government bookstore in your city, it may be quicker to obtain the book you want there, although there is no certainty they will have it in stock, or if they stock it, know where to find it.

The Food and Nutrition Board of the National Academy of Sciences' Institute of Medicine has the responsibility for compiling and issuing the *Food Chemicals Codex.* This reference includes physicochemical and microbiological specifications in monograph form for more than 900 food-grade ingredients approved for some sort of use in the US. Standards for the identity, strength, and purity of GRAS substances and of direct and indirect additives such as antioxidants, preservatives, flavors, sweeteners, nutrients, emulsifiers, enzymes, stabilizers, and colors are included in this volume.

Specifications contained in *Food Chemicals Codex* are frequently referred to in the *Code of Federal Regulations* (CFR) and they are accepted legal standards in Canada and are referenced in the food regulations of several other countries (e.g., Australia, New Zealand). The latest edition was issued in 1996. It can be purchased from National Academy Press, 2101 Constitution ave. NW, Lockbox 285, Washington DC, 20055, phone 1-800-624-6242, FAX 202-334-2451.

Computer Accessible Information

All federal regulations are contained in the CD-ROM from IHS Regulatory Products, a commercial firm (800-525-5539). This disk is said to include legislation in progress, Code of Federal Regulations, FSIS notices and directives, FDA guidelines and memos, speeches, background presentations, and directories. Frequency of updating is not known.

Chemical Abstracts becomes available on CD-ROM in 1996. Call Customer Service at 800-753-4227 for details.

A much larger and more diverse, but less dependable, resource is the Internet, a vast, chaotic, and confusing assemblage of information, disinformation, and blurry nothingness that constitutes the total body of international computer input into a an enormous pool of messages accessible with simple equipment and inexpensive software. The graphics-readable segment is the World Wide Web.

WWW sites — For those who are computer literate and who have equipment and line hookups that will enable them perform searches on the World Wide Web, a vast amount of information is available, some of it reliable and some not.

A searchable Code of Federal Regulations which contains all or virtually all of the national laws and regulations, including of course regulations related to foods and food ingredients, packaging, and labeling, can be found at the Internet address:

http://www.pls.com.8001/his/cfr.html

A variety of agricultural commodity outlooks, statistics, and other information is available at:

http://www.econ.ag.gov/

Other WWW sites that may have information of value for food scientists include:

American Cancer Society: **http://charlotte.npixi.net/acsfacts.html**
American Chemical Society, Publications Div.: **http://pubs.acs.org**
American Heart Association: **http://amhrt.org/ahawho.html**
American Medical Association: **http://www.ama-assn.org/**
Centers for Disease Control and Prevention: **http://www.cdc.gov/**
Department of Health and Human Services: **http://www.os.dhhs.gov/**
Food and Drug Administration: **http://www.fda.gov/**
National Cancer Institute: **http://www.nci.nih.gov/**
National Institutes of Health: **http://www.nih.gov/**
Natl. Inst. for Allergies & Infectious Diseases:**http://www.niald.nih.gov./**
National Library of Medicine: **http://www.nlm.nih.gov/**

The following sites allow searches that will locate a recipe by name or by key ingredients:

gopher://spinaltap.micro.umn.edu/1/fun/Recipes
file:://ftp.neosoft.com/pub/rec.food.recipes/
http://www.evansville.net/~wbbebout/food.html
http://www.spew.com/food/html
http:/www.epicurious.com/epicurious/recipes.html

Reliability — The WWW has attracted unscrupulous marketers, practical jokers, social activists, and the frankly psychotic; use it with care and skepticism. Larkin (1996) suggests using the following tests to estimate the reliability of a Web site:

•Who maintains the site? Government or university-run sites are among the best sources for scientifically sound health and medical information. Private practitioners or lay organizations may have marketing, social, or political agendas that can influence the type of material they offer on-site and which sites they link to.

•Is there an editorial board or another listing of the names and credentials of those responsible for preparing and reviewing the site's contents? Can these people be contacted by phone or through E-mail if visitors to the site have questions or want additional information?

•Does the site link to other sources of medical information? No reputable organization will position itself as the sole source of information on a particular health topic.

•When was the site last updated? Generally, the more current the site, the more likely it is to provide timely material. Ideally, health and medical sites will be updated weekly or monthly.

•Are informative graphics and multimedia files such as video or audio clips available? Such features can assist in clarifying medical conditions and procedures.

Search engines — Unlike libraries, the Internet contains no central indexing system. This anarchical state makes the collecting of all the available information on a specific topic very difficult, time-consuming, and problematical. Getting any information at all on an abstruse subject may be extremely tedious and, in the end, unfruitful. Search engines are tools that can help narow the field if you are looking for a specific topic, or if you have the name of an organization but no address for its Internet site. They allow you to input a few words that describe what you're looking for, and the searcher returns a list of sites related to your query.

Although a searcher can point the way to information, it cannot evaluate the information it locates. A typical search engine, tradenamed Webcrawler, is described as follows: "Webcrawler scans documents and counts the number of times a particular word or expression appears on a Web page. That alone determines whether the page is listed in our results and where it appears on the list." Unfortunately, this approach allows unscrupulous listers to attract viewers by including a large number of marginally pertinent, or even completely non-applicable, key words in their "page."

A few of the many search engines are:

Alta Vista: **http://www.altavisa.digital.com/**
Excite: **http://www.excite.com/**
Lycos: **http://www/lycos.com/**
Webcrawler: **http://www.webcrawler.com/**
Yahoo: **http://www.yahoo.com/Health/Medicine/**

chapter two

LOW-CALORIE AND NO-CALORIE SWEETENERS

Controlling body weight is a compelling concern of many Americans; most adults and many children are overweight and realize that they should attempt to become slimmer. Better appearance, better health, and improved quality of life are thought to be the rewards of weight reduction regimens for obese persons. Weight reduction can be achieved by increased energy expenditure, as by adhering to exercise programs, by reducing the quantity of food consumed, or by substituting lower calorie foods for those currently being consumed (the portion sizes remaining relatively unchanged). Of course, combinations of these approaches can also be effective.

The advantages of the slimmer look are apparent to most people, even those who prefer not to restrict their diet to achieve slimness. A significant segment of the population is willing to invest some effort in controlling weight. Recent studies indicate that between 33% and 40% of adult women and between 20% and 24% of adult men are dieting at any given time. For many individuals, a period of dieting is following by a period of weight gain. Typically, two-thirds of the weight lost during dieting is gained within one year of forgoing restrictions and almost all lost weight is regained within five years. Large and profitable industries have developed to cater to these cycling dieters.

Although excess sugar consumption has been blamed for the high level of obesity in America, the truth may be somewhat different. For at least twenty years it has been conventional wisdom that sugar consumption represents about 30% of the daily caloric intake, an estimate based on the amount of sugar delivered into the food supply ("disappearance" as tabulated by the USDA). The FDA has, however, recently estimated that sugar consumption among US residents constituted only 11% of the total dietary calories. Even so, a reduction in the consumption of sugar and other utilizable carbohydrates is useful in achieving weight loss.

USING ALTERNATIVE SWEETENERS TO OPTIMIZE BODY WEIGHT

It has been recognized for a rather long time that weight reduction regimens would be greatly facilitated if sugar could be replaced by a substance that would give products of acceptable sweetness but with much lower calorie content than similar products made with sugar. As a result, there has been a great deal of experimentation and marketing effort spent in the field of alternative sweeteners.

According to Gelardi (1987), the ideal alternative sweetener would be as sweet as (or sweeter than) sucrose, it would have a pleasant taste with no aftertaste, and it would be colorless, odorless, readily soluble, stable, functional, and economically feasible. "It also is nontoxic, does not promote dental cavities, and is either metabolized normally or excreted from the body unchanged without contributing to any metabolic abnormalities." As Gelardi puts it, however, the ideal, all-purpose, low-calorie sweetener simply does not exist. A lot of time, effort, and money has been spent in trying to approach that goal. A few of the more successful alternative sweeteners will be reviewed in this chapter.

ALTERNATIVE SWEETENERS

Carbohydrates are used for other reasons besides sweetening, but the flavor contributions of sugars is a critical property justifying the inclusion of these calorically important ingredients in many foods. This section will discuss not only those sweeteners that contribute no nutritive value to the diet, such as saccharin, but also high-intensity sweeteners that may contribute a very minor amount of physiologically utilizable calories, such as aspartame. Some writers prefer to call these replacements "alternative sweeteners" (Giese 1993).

There are several characteristics on which sweetener suitability for a given application can be judged. One list includes regulatory status, taste, solubility, stability, and cost. Another list of properties that sweeteners may or may not contribute, depending on the type of sweetener and its concentration in the product, includes:

1. Provides inexpensive bulk (weight and/or volume).
2. Provides a universally liked taste.
3. Increases viscosity to improve mouthfeel of beverages
4. Has agreeable physiological effects.
5. Interacts with other constituents (e.g., pectin) to provide structure.
6. Acts as fermentation substrate.
7. Functions as a preservative when present in high concentration.
8. Controls starch gelatinization.
9. Modifies crumb tenderness in cakes and other bakery products.
10. Reacts with proteins and amino acids to provide flavors and colors.
11. Mellows acidic and bitter flavors.
12. Increases water activity and humectancy.
13. Provides the entire structure of hard candy and the like.

Failure to take all of these factors into consideration when developing low calorie or no-sugar-added products can lead to very unpleasant surprises when the formula reaches the production floor and when the product goes into distribution.

Alternative sweeteners can be classified into two groups: those synthesized from industrial chemicals and those extracted from plants. In many cases, the sweeteners of plant origin are modified somewhat by chemical means to give the substances offered as food additives.

Synthesized Chemicals

The state of the art of chemical modification has advanced to the point where preparation of high intensity sweeteners is not a particularly difficult endeavor, but the obtaining of substances that have no appreciable side effects and that can be approved by regulatory agencies leads to almost insuperable problems. As a result, only a very few artificial sweeteners can currently be used in any kind of food or beverage.

Acesulfame-K — This sweetener was approved in 1988 for dry beverage mixes, tabletop sweeteners, gelatin desserts, chewing gum, instant coffee and tea, nondairy creamers, and puddings. In 1994, it was approved for use in some bakery foods. It is a dipeptide of aspartic acid and D-alanine linked to an unusual cyclic amide, in chemical terms being the potassium salt of 6-methyl-1,2,3-oxathiazine-4(3H)-one-2,2-dioxide. Several foreign countries approved the substance for food use before the FDA took this step, and many low-calorie products made outside the U.S contain the compound. A tradename is "Sunette."

Acesulfame-K is a white, odorless, readily soluble crystalline substance about 200 times as sweet as sucrose. It is said to be very stable in both dry and dissolved forms and it is not metabolized in the human body. The FDA has approved the use of an encapsulated acesulfame-K.

Bullock et al. (1992) described experiments in which acesulfame replaced sugar in cookie doughs and the effects of three different bulking agents (polydextrose, powdered cellulose, and soy fiber) were compared.

Formula

37.12 parts all-purpose flour
24.43 parts hydrogenated vegetable shortening
15.68 parts pasteurized whole eggs
10.45 parts water
 1.35 parts flavorings
 0.48 parts salt
 0.37 parts baking soda
 0.20 parts Acesulfame-K
10.00 parts polydextrose, or 5.00 parts powdered cellulose, or 5.00 parts soy fiber.

Procedure

Shortening and flavorings were creamed in a standard bench-top size vertical dough mixer. Dry ingredients were mixed in a separate bowl, then added. Liquids and dry ingredients (which had been pre-mixed in a separate bowl) were added in alternate portions to the mixer bowl, the agitator being operated intermitently. Total mixing time was five minutes. One ounce portions of the dough were baked in a convection oven.

Results

The dough did not spread during baking, which could probably be regarded as a negative, depending on the results intended. Cookies made with soy fiber required the least force to break (were the tenderest), followed by powdered cellulose cookies, and polydextrose cookies. The cookies sweetened by Acesulfame-k were described by taste panelists as being softer and more cake-like than cookies prepared with sugar (which were crisper). Nutrient values (mostly calculated using database values for ingredients) are shown in Table 2.1.

Table 2.1
NUTRIENTS IN COOKIE DOUGHS MADE WITH ACESULFAME-K[1]

Constituent	Sucrose control	Poly-dextrose	Cellulose	Soy fiber
Calories, kcal	116.0	109.0	112.0	113.0
Moisture, %	24.0	27.9	29.3	29.8
Protein, %	5.6	5.7	6.0	6.7
Carbohydrate, %	47.1	28.3	29.8	29.8
Fat, %	21.8	26.4	27.8	30.9

Source: Bullock et al. (1992).
[1] "No added sugar" versions plus control. Calculated from database.

As can be seen from the composition, there can be no substantial reduction in calorie count in the artificially sweetened samples. These formulas might be used as a basis for no-sugar added or reduced-carbo-hydrate cookies, but they could not be labeled as a food for use in reduced calorie diets. When rated on a 7-point hedonic scale for overall acceptability, the cookies made with polydextrose received a rating of 4.4, while those made with cellulose fiber and the soy fiber cookies both rated 2.8. As the authors explained, the formulas were not intended to be in final form at the end of this series of experiments.

Alitame — This compound is a dipeptide-type sweetener marketed under the trade name "Aclame". It is made by chemical synthesis and contains the amino acids L-aspartic acid and D-alanine,and a novel amide substituent. The chemical formula is L-α-aspartyl-N-(2,2,4,4-tetramethyl-3-thietanyl)-D-alaninamide.

Alitame is said to be 2,000 to 3,000 times sweeter than sugar (Anon. 1994E); it is partially metabolized, yielding about 1.4 calories per gram.

The manufacturer describes the taste as clean, sweet, and sugar-like. The compound is sufficiently heat-stable to be useful in confectionery and thermally pasteurized foods, and in high-temperature processed foods of neutral pH (such as bakery foods). A slight amount of decomposition may occur, however. FDA approval is pending.

Freeman (1989) published formulas for several kinds of cakes sweetened partially with alitame. The formula for pound cake may be of interest. It is reproduced below, with minor changes (percentages apply to batter composition, as is).

Formula — Part A

12.0% crystalline sorbitol
2.5% maltodextrin
5.5% N-flate (a proprietary blend of emulsifiers and other ingredients)
0.15% sodium bicarbonate
0.165% total of butter flavor, vanilla flavor, and lemon flavor.
0.008% alitame
1.0% cellulose
0.11% salt

Formula — Part B

23.5% cake flour
0.35% sodium aluminum phosphate
0.2% sodium bicarbonate
10.0% polydextrose

Formula — Part C

23.72% whole eggs
12.3% water
0.001% FD&C Yellow 5

Formula — Part D

8.5% water

Preparation Method

Blend part A. Slowly add part B, and mix. Add part C, mix, scrape down the bowl, and mix until smooth. Add part D and mix until smooth. Scale 14 oz batter into a loaf pan with 8x4 inch top, 7x3 inch bottom, and 2.5 inch deep; greased or paper-lined. Bake at 350°F for 35 min.

Results

The finished products had acceptability comparable to cakes sweetened with sugar.

Aspartame — This compound (tradenamed "Nutrasweet") does have some calorie content, but the energy value it contributes to finished products is negligible in most applications. Aspartame is prepared by chemical synthesis and is the methyl ester of 1-aspartyl-1-phenylalanine. The compound is said to taste about 180 times as sweet as sucrose, but its relative sweetening power depends on the conditions under which the two compounds are compared.

Although it is only about a half to a third as sweet as saccharin and is more expensive than that substance on a weight basis, aspartame does not contribute the bitter and metallic notes frequently observed when foods and beverages containing saccharin are ingested. When it is mixed with acidic aqueous sytems (e.g., carbonated beverages), aspartame is slowly hydrolyzed and loses its sweetening power; this reaction is accelerated by high temperatures.

The FDA has approved an encapsulated form of aspartame, which is said to have improved resistance to heat.

Cyclamate — This sweetener was discovered in 1937, and became quite popular in the 1960s as a beverage sweetener in combination with saccharin. Although it is only about 30 times as sweet as sugar and was more expensive than saccharin, it was used in combination with the latter compound in order to make saccharin's bitter taste less noticeable. It was at the time supposed to be GRAS, but the FDA terminated this clearance in 1969 until additional safety data could be submitted by the manufacturer. Data acceptable to the FDA was never submitted, and the sweetener is not now in use in the US. The regulatory history of this substance is an interesting study of bureaucracy in action.

L-sugars — L-sugars are stereoisomers of the common and natural D-sugars such as D-glucose. Although these materials stimulate the taste buds, the human body cannot metabolize them and so they contribute sweetness without calories.

Neosugar — Neosugar is a derivative of sucrose produced by a fungal enzyme and is non-metabolizable, thus being a no-calorie sweetener.

Saccharin — The chemical called "saccharin" is 2,3-dihydro-3-oxobenzisosulfonazole. It was discovered over 100 years ago. Saccharin is about 300 to 500 times sweeter than sucrose when the two substances are compared in dilute aqueous solutions. An exact match to sucrose cannot be obtained at any concentration because saccharin has flavor notes that are not present in sugar, and vice versa. Many people find that saccharin has bitter, astringent, or metallic tastes, especially at high concentrations.

These off-flavors tend to be particularly objectionable in delicately flavored fruit products, but may be partly concealed in foods that contain small amounts of sucrose or corn sweeteners in addition to saccharin, and they are also less obvious in cola beverages, hot cocoa and other chocolate products, coffee, etc.

Saccharin is very stable under conditions normally encountered in the preparation and storage of foods and beverages. It is inexpensive compared to most other artificial sweeteners.

The advantages of cost effectiveness and stability have led to continued wide use of saccharin in spite of its recognized flavor disadvantages. It is currently being used in soft drinks, fruit juice drinks and other beverages, processed fruits, chewing gum and confections, gelatin desserts, jams and jellies, salad dressings, and baked goods, and as a table-sweetener.

Sucralose — Sucralose is a derivative of sucrose in which three atoms of chlorine have been substituted for three hydroxyl groups on the sucrose molecule, yielding a substance having the chemical formula 4,1'6'-trichloro-4,1,6'-trideoxy-galactosucrose.

Sucralose is said to be about 550 to 750 times sweeter than sugar and to have zero caloric value (Hood and Campbell 1990). Solubility in water increases from from 25% (w/w) at room temperature to 68% at 78°C. It has considerable heat stability and has viscosity characteristics (in water) similar to sugar.

Radioactively tagged sucralose was used as an ingredient in cakes, sugar cookies, and graham crackers and the finished products extracted to determine the deterioration products (Barndt and Jackson 1990), It appeared that little or no breakdown of sucralose occurred under the conditions of baking, indicating the excellent stability of this sweetener.

Sucralose is still under FDA consideration. A trade name used by McNeil Specialties is "Splenda"

Formulas and processing methods for the experimental products containing sucralose studied by Hood and Campbell are given in the following tables which include a type of sugar cookie, chemically leavened yellow cake, a chocolate cake mix, and oatmeal cookies. The original publication lists formulas for apple pie fillings and some other bakery type items.

Cookies with Sucralose
(Hood and Campbell 1990)

	%
Flour	45.3
Whole eggs	2.0
Shortening	10.0
Salt	0.44
Water	11.7
Baking powder	0.52
Polydextrose	25.0
Dextrose	2.0
N-Flate (tradename)	3.0
Sucralose	0.03

Preparation method: Blend all dry ingredients except sucralose. Dissolve sucralose in the water, then add the liquid to the dry ingredients. Mix for about 1.5 minutes. Roll out the dough into a thin sheet and cut into uniform discs weighing about 7 grams each. Bake for 12 minutes at 400°F.

Yellow Cake Made with Sucralose
(Hood and Campbell 1990)

	%
Flour, cake	25.84
Water	26.94
Polydextrose	13.75
Egg white	12.37
Shortening, emulsifier	5.50
Whole egg, fresh	4.12
Fructose, crystalline	2.75
Starch, pregel	1.65
Emulsifier (2.6 HLB)	1.65
Non-fat dry milk	1.38
Baking powder, double acting	1.38
Baking powder, slow acting	1.38
Salt	0.55
Flavors	0.36
Lecithin	0.16
Sucralose	0.11
Preservative	0.10
Color	0.01

Preparation method: The yellow cake formula performed well on scale-up testing in an Oakes continuous mixer. Details of the preparation were not given, bu presumably included the usual steps of preparing a dry mix followed by liquid addition, then thorough mixing short of any developemnt.

Results: The reduced calorie product demonstrated essential equivalence in flavor, texture, and overall acceptability with cakes made from a control formula that contained most of the ingredients in the test cake, except for no polydextrose and sugar and increased amounts of shortening and eggs. Calculated caloric content showed a 36% reduction for the low-calorie cake, on an equal serving weight basis.

Chocolate Cake Mix with Sucralose
(Hood and Campbell 1990)

	%
Flour, cake	19.03
Water	43.01
Cocoa, 10-12% fat	4.28
Litesse, a fat replacer	4.76
Shortening, emulsifier	7.23
Whole egg, liquid whole	9.51
Fructose, crystalline	4.76
Wheat starch	2.85
Non-fat dry milk	0.95
Baking powder, double acting	2.66
N-flate (proprietary improver)	0.38
Salt	0.38
Carboxymethyl cellulose	0.14
Sucralose, liquid concentrate	0.20

The chocolate cake mix is said to furnish a finished cake having 220 calories in a 100 gram serving as compared to about 347 for a representative sugar-sweetened cake. Some of the calorie reduction comes from reductions in fat, flour, and sugar as compared to a normal cake recipe. In the test product, 56% of the calories are from carbohydrate, 37% from fat, and 7% from protein.

Pies in which the fillings were sweetened with sucralose were also described. The formula and processing method given were generally of the conventional pattern, the filling being made of 60.00% IQF apples, 36.30% water, 2.50% pregel starch, 0.80% flavors, and 0.40% of a 10% sucralose solution. The writers said that apple pie filling had outstanding sensory qualities, but no other details were given. A reduction of 33% in calories as compared to a control made with conventional formulas was determined by calculation. This figure must apply to the filling alone, since the crust, according to the formula shown here, would contribute considerably more calories than the filling, and the crust's formula was not modified by artificial sweetener addition.

Oatmeal Cookie Made with Sucralose
(Hood and Campbell 1990)

	%
Flour	25.25
Polydextrose	20.28
Oats	13.44
Water	16.17
Raisins	7.61
Shortening	4.44
Corn syrup	3.11
Molasses	3.11
Nonfat dry milk	2.03
Fructose, crystalline	1.27
Baking soda	0.98
Spices/flavors	0.74
Salt	0.63
Baking powder	0.41
Sucralose, 10% solution	0.38
Glycerol	0.05

Details of mixing, forming, and baking the oatmeal cookies were not given by the authors, so we can reasonable expect that they followed a standard type of procedure for making soft and chewy cookies. In taste tests, the reduced calorie product "almost matched the control commercial product." Calorie count was reduced by 33% from the control.

According to the patent of Beyts (1990), and perhaps others, beverages such as carbonated, acid-pH soft drinks, and tea and coffee, can be improved if a combination of two sweeteners is used; for example, a chlorosucrose such as sucralose and cyclamate. A formula for a cola drink representing the claims of the patent is given:
0.0154% phosphoric acid (88%)
0.0154% sodium benzoate
0.48% cola extract
0.1% cola essence
0.0046% sucralose
0.0613% cyclamate
To the preceding ingredients, carbonated water is added to make 100%

Tagatose — Tagatose is a crystalline sugar of the ketohexose class. It has been known for many years, the dextro variety being originally prepared by treating galactose with dilute alkali. Apparently it is not utilized by humans, so it is physiologically "non-caloric." Of course, in bomb calorimetry, it would yield about the same number of calories as glucose, fructose, galactose, etc.

Patents have been secured by Biospherics, Inc. on methods for making, purifying, and utilizing D-tagatose as a food ingredient. To manufacture this sugar, whey permeate is enzymatically hydrolyzed to galactose and glucose. The galactose is separated and purified, then isomerized to tagatose by alkalizing with lime. The calcium tagitate solution that results is acidified, concentrated, and dried to yield tagatose crystals of fine particle size.

Tagatose is said to be about 92% as sweet as sugar (Kevin 1996), and apparently resembles sucrose in many of its physical properties so that reformulation of existing sugar-sweetened foods can be greatly simplified. The attitude of the FDA toward use of the material in foods is not clear, but it has evidently not been approved. Biospherics, Inc. may seek GRAS clearance, since the sugar has been reported as occurring naturally in some cheeses, yoghurt, cocoa, dates, and heat-treated milk products.

High Intensity Sweeteners From Natural Sources

There are several low-calorie or no-calorie, high intensity sweeteners derived from microorganisms or plants (or chemically modified materials from these sources) that seem to have some potential for dietetic foods but have not become widely accepted in the US, in some cases because they have not received FDA clearance, in other cases because there are negative aspects connected with their functionality — such as off-flavors, poor stability, questionable supply, high cost, or incompatibility with other ingredients. They often have curious sensory side-effects such as delayed impact on the taste buds or off-flavors that linger or change in character with time. Some of these materials have been under investigation for decades. Prominent examples include neohesperidin, glycyrrhizin, stevioside, monellin, miraculin, and thaumatin.

Erythritol — This is a tetrahydric alcohol originally isolated from certain lichens and algae. It can also be obtained from the decomposition of erythrine, an alkaloid obtained from certain species of the genus *Erythrina*, which includes particular kinds of shrubs and trees related to the pea family. It can also be made synthetically.

The purified substance is a colorless, sweet, crystalline material that occurs in different optical varieties; in nature, it is found in the meso form.

Apparently, it is not utilized in the human body, justifying a non-caloric claim. It is manufactured in Japan, where it is used as a sweetener in sugarless chewing gum and perhaps in other products. Approval for use in foods in the US has not been granted, but Tyrpin et al. (1996) received a patent covering use of erythritol in chewing gum as a non-caloric, non-cariogenic sweetener.

Glycyrrhizin — This substance is a non-caloric extract of liquorice (also spelled licorice) root. Stick licorice, a partially dried extract of the root having a very sweet and somewhat astringent taste, is an old item of commerce, having been used as a component of confections for well over a hundred years. It has also been used in liqueurs. Glycyrrhizin, a colorless crystalline glucoside is the compound mainly responsible for the sweet sensation (*Glycyrrhiza* is the genus in which the licorice plant is classified).

Ammoniated glycyrrhizin is the form of the sweetener receiving the most publicity and promotion as a high intensity, low or no calorie sugar substitute. It is said to be 100 times sweeter than sugar. Unfortunately, ammoniated glycyrrhizin's strong licorice flavor appears to be inseparable from the sweetening effect, which greatly limits the ingredient's use in most food products. This substance has been applied mostly in tobacco and pharmaceutical manufacture, but a few confectionery products containing it have been made. Glycyrrhizin has foam-enhancing properties, which might be useful in beverage formulation.

Miraculin — The berry of the African miracle fruit *Richardella dulcifie* contains this proteinaceous substance. Miraculin is "2,000-2,500 times sweeter than an 8 to 10% solution of sucrose" (Dziezak 1986). Reportedly, it has other properties as well, including masking of sour taste, flavor enhancement, and various synergistic properties when combined with other flavors (Anon. 1996B). It is known to have good heat stability and resists change by pHs as low as 2. Miraculin was denied GRAS status by the FDA, but has been introduced into Jamaica, Puerto Rico, Brazil, and Panama (Newsome 1986).

Monellin — The West African serendipity berry, the fruit of *Discoreophyllum cumminsii*, yields this sweetener, which was heavily promoted by a US research foundation. Apparently, instability problems and other difficulties have precluded its approval and use in this country.

Neohesperidin — There is a family of products (dihydrochalcones) derived from raw materials (flavanones) found in the peels of citrus fruits, especially naringin from grapefruit. Some of these have very high sweetening power, being 1,000 times (or more) sweeter than sucrose.

Stevioside — This sweetener has been used for centuries by the natives in Paraguay. It is obtained from the roots, stems, and leaves of a South American herb, *Stevia rebaudiana*. It is about 300 times sweeter than sucrose, and has a somewhat bitter aftertaste. The stability of the substance to processing conditions and storage is generally good. It has not been approved in the US but is widely used in Japan, and perhaps in other countries.

Thaumatin — The flavor preparations called "thaumatin" are based on proteins of low-molecular weight extracted from the berry of a West African plant, *Thaumatococcus danielli*. These materials are 750 to 1,600 times sweeter than sucrose, and they have an aftertaste reminiscent of anise or licorice. They are relatively stable in solution and to heating, and have been used as flavor enhancers and non-nutritive sweeteners alone or in combination with saccharin and acesulfame-K. Australia, Japan, and the UK permit use of thaumatin as a non-nutritive sweetener, but in the US it is allowed to be used only as a flavor adjunct in chewing gum. A trade name, Talin, is used for the ingredient most publicized in the US.

chapter three

"DIABETIC" OR SUGAR-FREE FOODS

Diabetics have a more serious problem with sugar than do the weight-loss dieters. To diabetics, sugar may indeed be a poison, and some of them may be unable to tolerate even traces of the sweetener. This extreme sensitivity of the consumer, and the responsibility placed on the vendor of foods labeled as being "for diabetics," place heavy burdens on the designer of these foods.

SIMPLE AND COMPLEX CARBOHYDRATES

The fallacy of accepting conventional wisdom on solutions to nutritional problems, especially those dogmas based on non-scientific extrapolation of the results of investigations yielding inadequate or poorly analyzed data or having unsuitable controls, or on demographic studies based on mixed and poorly defined populations, is clearly illustrated by the history of lists of carbohydrate substitution that for many years have been followed by diabetics on the advice of their physicians. The following discussion, based in part on an article by Kolata (1983), describes that problem.

Until about the middle 1980s, diabetologists and nutritionists taught that there are two major classes of carbohydrates: simple and complex. Their theory stated that simple carbohydrates, which are the mono- and disacchrides such as glucose, sucrose, and fructose, are immediately absorbed by from the intestine, rapidly enter the circulatory system, and cause a rapid rise in blood sugar and blood insulin. Complex carbohydrates, such as the starches found in rice and potatoes, take longer to be absorbed and so result in a slower and more moderate rise in blood glucose and blood insulin. Therefore, the theory stated, complex carbohydrates were much less harmful to diabetics than were simple carbohydrates, and, therefore, were less effective in causing diabetic symptoms to develop in susceptible persons.

As it happens, this theory is incorrect. The lists or tables of carbohydrate equivalents followed religiously by so many patients for decades were essentially of no value in improving or maintaining their health or quality of life. A diabetes specialist at the National Institutes of Health said, "I believed it. Everyone believed it. But no one ever tested it."

Finally, someone did test it. Phyllis Crapo of the University of Colorado Health Sciences Center in Denver thought it would be worthwhile to subject the conventional wisdom to experimentation, and she was astonished to find it had little basis in fact. Crapo and other researchers learned, for example, that a bowl of ice cream has little effect on blood glucose, and neither does a sweet potato. On the other hand, a white potato or a slice of whole wheat or white bread markedly elevates blood glucose. To further complicate the matter, effects of carbohydrates on blood glucose are difficult to predict based on chemical characterizations or other readily determinable qualities. The only sure way to establish the effect of a particular food on blood glucose levels is to test it on volunteers.

These discoveries were of major importance for diabetics, whose health and life depend on avoiding large fluctuations in blood glucose. Even for nondiabetics this policy may be worthwhile, since large amounts of glucose in the blood make people sleepy and may have other undesired results.

Other effects are more speculative, but some medical researchers have suggested that the reason people develop adult-onset diabetes may be because they eat the wrong kind of carbohydrates, meaning those that give rapid rises in blood sugar, rather than simply consuming too many carbohydrates. In more primitive societies, David Jenkins of the U. of Toronto points out, people tend to eat the carbohydrates that give slow glucose rises and tend also to have little diabetes and heart disease.

Even before the crucial tests were made, it was recognized by many physiologists that the traditional advice to diabetics that they eat complex rather than simple carbohydrates really made no sense. People have so ample a supply of amylases in their digestive systems, that starch and most other so-called complex carbohydrates are quickly converted to glucose. Of course, this is not the entire answer, either, since it has become evident in recent years that some forms of starch are particularly resistant to amylases.

As a nutritionist, Crapo had been advising diabetic patients to concentrate on complex carbohydrates and she found it hard to believe that they are no different from simple sugars as far as blood glucose is concerned. So she decided to feed volunteers pure uncooked starch and see what happened. As she suspected, there was virtually no rise in blood glucose.

Investigators were aware that pure uncooked starch is hardly a typical food. Perhaps the result would be different if they fed volunteers cooked starches of the sort people normally eat, starting with potatoes and rice. To their amazement, says Crapo, "We found a dramatic difference between the two. Rice gave a flat glucose response and potatoes gave a rapid response that was the same as you would expect if you gave people pure glucose. "Potatoes are like candy as far as a diabetic is concerned."

Next, they tested corn and bread because rice, potatoes, corn, and bread provide the major amount of starch eaten in this country. The ranking of these foods in order of increasing effect on elevation of blood glucose levels is rice, bread, corn, potatoes.

They tried the experiments in people with impaired glucose tolerance and with diabetes. The results were the same. They tried similar experiments using simple sugars because it was generally believed that all simple sugars reacted the same way. It turned out that simple sugars are as different as potatoes and rice. Lactose and fructose have little effect on blood glucose. Sucrose has a moderate effect. Glucose and maltose give immediate and pronounced effects.

One investigator hypothesized that the reason for these results is that there are differences in the accessibility to digestive action of the starch or sugar molecules in various foods. The more homogenized the food, the more rapid the rise in blood glucose. A rice slurry gives a more rapid rise than a whole apple. The biochemistry of food digestion and absorption is so poorly understood, he said, that each food has to be analyzed separately.

An extensive list of the glucose responses of foods was prepared by Jenkins and his associates in Toronto collaborating with Thomas Wolever and his staff at Oxford U. For example, they found that legumes caused blood glucose increases about half those of their cereal counterparts. Pasta products are said to be much lower in their blood glucose effects than cereals. There is, however, little or no difference in blood glucose levels with white or whole wheat pasta, white or whole wheat bread, or white or brown rice Jenkins and Wolever also tested combinations of foods; a cheese-bread combination gives the rapid glucose rise characteristics of bread. Bread and beans, in contrast, resulted in a slow rise in blood sugar more characteristic of beans than bread. Corn syrup, in spite of being mostly glucose, gives a very slow blood glucose response whereas sucrose gives a more rapid one.

One consequence of these results for diabetics is that the starch exchange tables used in managing their diets, which are lists of equal carbohydrate portions of (for example) bread, rice, potatoes, or corn, were called into question. In addition, Olefsky points out, diabetics may want to know that some foods that they frequently avoid, such as ice cream, are fine as far as blood glucose is concerned. Olefsky remarks that he recently saw a new ice cream sweetened with sorbitol and labeled, "not a low calorie food." The ice cream was aimed at diabetics, but regular ice cream gives a very flat glucose response.

BREAD WITHOUT SUGAR

There seems to be a demand for "sugar-free" bread by diabetics and other persons who believe that sugar is a very harmful substance. The term "sugar-free" can mean different things to different people. Some health food buyers wish to avoid refined white sugar (meaning the ingredients called sugar, beet sugar, or cane sugar), but would probably accept a product that contained honey, molasses, corn syrup, malt syrup, or even the so-called "raw" or "unrefined" sugars. Few production problems will be encountered when one of these ingredients is substituted for the small amount of sugar usually added to bread dough, but flavor and color may be affected significantly.

Even if sugar-free is interpreted to mean "does not contain added sucrose, glucose, or fructose," the demand can be met provided the baker is willing to make minor modifications in the standard bread formula. Corn syrup, honey, molasses, and sweeteners prepared from fruit juices should not be used in bread described as "free of added sugar(s)."

Formulas for regular white bread contain sugar in amounts too small to have much influence on the health of diabetics, assuming the usual consumption of a couple of ounces of bread per meal. Most of the carbohydrate content of bread is in the form of starch. Some glucose and maltose is formed in the dough after water is added, provided amylases are present, but only a small quantity of sugar remains at the end of a normal fermentation schedule. The regulatory status of mono- and di-saccharides created during dough processing is not clear.

It is assumed diabetics will receive dietary guidance from their physician and that this will include advice on whether to moderate or cease their consumption of bread and rolls. Sweet baked goods (cakes, pies, pastries, etc.) will probably be described by the medical adviser as being unsuitable for a diabetic's diet.

Sugar itself does not play an important role in establishing the structure of bread dough, but it has an indirect influence on the physical properties of bread as a result of its function as a fermentation substrate. Since sugar is usually added at the level of only a few percent in white bread and is rapidly converted by yeast into glucose and fructose which are then metabolized by the cells, the amount of sugar present in dough is continually reduced as the dough ferments.

As a source of preferred yeast nutrients, sugar's effect on fermentation is substantial but it can be replaced to some extent by the glucose and maltose generated by diastatic malt syrup or fungal amylases. Malt syrup, whether diastatic or non-diastatic, includes substantial percentages of glucose, maltose, and higher saccharides, and these constituents should be taken into account when making labeling claims. It is obvious that bread from which all sugars have been omitted or consumed by yeast will be less sweet than regular white bread and, therefore, less palatable to most consumers.

Alternative sweeteners that allow bakers to claim either sugar-free or no sugar added while still retaining the approximate volume and weight of a conventionally formulated food include sugar alcohols, hydrogenated starch hydrolysate, isomalt, lactitol, maltitol, mannitol, sorbitol, polydextrose, and xylitol. The response of the diabetic consumer to these substances is not always clear-cut. All of them have sweetness levels different than sucrose, usually lower, and some of them have flavor notes significantly different from sucrose. The sweetness can be supplemented with one of the high intensity sweeteners, such as saccharin, but masking the resulting off-taste may be difficult, especially in delicately flavored foods.

Milk products (except caseinates) include considerable amounts of lactose, a disaccharide sugar. Common ingredients that include this sugar are milk, skim milk, whey, cream, buttermilk,and yoghurt, whether dried or liquid and whether treated with the enzyme lactase or not. Butter, and preparations containing butter, will probably contain a small amount of lactose in the water phase that constitutes 15% to 20% of these compounds, and this also applies to those varieties of margarine which include some skim milk as an ingredient. It is reasonably clear that products containing any of these ingredients cannot be truthfully called "sugar-free." Caseinates and purified butterfat should be free of lactose, however.

As an alternative approach that eliminates the need for fermentation and the sugar that is required for that process, simulated bread loaves can be made by a chemical-leavening process. Since these loaves will be devoid of the flavors developed during fermentation, most consumers will quickly identify them as being foreign to their experience, and therefore, usually, less desirable than conventional bread and rolls.

BULKING AGENTS TO REPLACE SUGAR

When sugar, corn syrup, starch, and other metabolizable carbohydrates are removed from a formula, a weight and volume void is left which, in many cases,must be filled, not only for the marketing need to furnish a package or portion resembling the "normal" food in appearance and size, but also for texture and other consumer-satisfaction characteristics. Obviously, it would do no good to replace the missing carbohydrate with protein or fat, so what is needed is an edible substance contributing few if any calories that can make up the volume and weight of the missing sugar. Powdered cellulose is certainly one possibility, but it does have disadvantages for many kinds of foods; labeling declarations for example. There has been an extensive research effort to seek out natural bulking agents and to synthesize others.

The problem becomes particularly acute, when a large amount of sugar is replaced with a small amount of high-intensity sweetener. In these cases, the product designer is left with a finished food that has a batch weight and volume, and, often, a portion weight and volume, much smaller than those possessed by the unmodified product. In formulating beverages, the difference can be made up by water or whatever other liquid constitutes the bulk of the beverage. Powders which the consumer mixes with water to form beverages also present relatively simple problems. In some cases, the sugar also contributes to the texture of the finished product, as in pectin-based jellies where the strength of the finished gel is strongly influenced by the sugar concentration, and in icings, where the actual strength and form of the material is based on sugar. Addition of water, in the absence of sugar, can have serious effects on the shelf-life of many those products in which the sweetener contributes significantly to spoilage protection.

Several types of candies are good examples of the structural effect and preservative effects of sugar, In these, sugar makes up a substantial portion of the product's weight and/or volume or forms the basic structure of the product. As one of the simplest examples of this problem, consider hard candy, such as Lifesavers. Some of these products contain about 95% sugar. If this sugar, or a substantial percentage of it, is replaced by a very small amount of high-intensity sugar-replacer, the finished product appears totally different and the portion size becomes so small as to be ridiculous. Similar problems, only slightly less severe, occur when making meringues, and in many cakes and cookies. What is needed, then, is a non-caloric bulking agent that can replace the volume and weight contributions of the missing sugar and, if possible, bring the water activity to about the same level found in the conventional product.

Suppliers have recognized this need, and various non-caloric or low-caloric bulking agents have been made available to formulators of dietetic foods. Among them are sugar alcohols, hydrogenated starch hydrolysates, isomalt, lactitol, maltitol, mannitol sorbitol, and xylitol; also polydextrose. All of them seem to be more expensive than cane or beet sugar, and some of them contribute flavors not expected by the consumer and generally not liked in their unfamiliar context. A few of the candidates will be discussed in the following paragraphs.

Maltodextrins

These materials are probably the most widely used bulking agents in foods. They are relatively inexpensive, and have not been accused of causing any health problems. Maltodextrins are fully digestible and have approximately the same caloric density as dried corn syrups, so they cannot be claimed to be "low-calorie," but they do have the advantage for many applications of having little or no sweet taste.

Maltodextrins are produced in much the same way as corn syrups, i.e., by enzymic breakdown of corn starch, but conditions of the hydrolysis are adjusted so that the dextrose equivalent of the finished product is less than 20, some have DEs as low as 4 or 5. The higher DE products have better cold-water solubility and more sweetness, contribute more to the browning reaction, and have a higher level of hygroscopicity. Low DE products add more viscosity, increase moisture retention, and have almost undetectable sweetness. Corn maltodextrin gives a clear solution while rice maltodextrins generally have some opacity, partly because they contain detectable amounts of protein and fat.

The dried syrup, which is the commercial ingredient, is used to increase bulk and/or soluble solids, inhibit sugar crystallization, reduce freezing points, serve as carriers for flavors, colors, etc., and increase viscosity. These substances are claimed as being GRAS.

There are several versions of maltodextrins, each designed to have some particular property that is of interest to a specific segment of the food industry. As an example, maltodextrins of very low bulk density enable the manufacturer to give the customer the container volume and serving volume he expects, without having to contend with useless weight and extra calories.

Polyols

Polyols, some of which are called sugar alcohols, offer various combinations of dietetic advantages with sweetness (usually less than sugar), humectancy (usually greater than sugar), low or no participation in Maillard reactions, substantial freezing point depressions, and variable chelating properties. There are several varieties of these materials; many of them can be manufactured by catalytic hydrogenation of sugars and oligosaccharides. They have varying degrees of dietary energy content, often substantially less than sugar's four calories per gram, and in a few cases approaching zero.

Many polyols have been promoted as being much less cariogenic than sugar, glucose, fructose, etc. That is, candies made with polyols as their major ingredient are said to be less conducive to tooth decay than products made with sugars.

Calorie content of sorbitol must be declared on the label at 2.6, maltitol at 3, and mannitol at 1.6. It is not clear whether these polyols should be declared as sugars, the manufacturer says they need not be. Hydrogenated starch hydrolysates (made by hydrogenating maltose corn syrup) are said to contribute 3 dietary kcal/gram, xylitol at 2.5, lactitol at 2.0, and isomalt/palatinit at 2.0.

Hydrogenated starch hydrolysates — Noncrystallizing syrups containing sugar alcohols and other materials can be manufactured by contacting corn syrups with hydrogen gas in the presence of a catalyst. The generic name for these products is "hydrogenated starch hydrolysates." They contain a mixture of compounds, their composition varying according to both the conditions of hydrolysis and the method of hydrogenation. Various products will contain different amounts of maltitol and sorbitol. Petitions for GRAS status have been filed. Sweetness and actual caloric contributions to the human diet will vary, but most types would probbly rate close to four calories per gram.

Isomalt — This bulking agent, one version of which has been tradenamed "Palatinit," is made by hydrogenating isomaltulose in an aqueous solution using a metallic catalyst; principal constituents are α-D-glucopyranosyl-1,1-D-mannitol and α-D-glucopyranosyl-1,6-D-sorbitol in about equal amounts (Irwin 1990). The raw material, isomaltulose, is obtained by the transglucosidation of sucrose, using a mutase enzyme extracted from cultures of *Protaminobacter rubrum.*

Isomalt is said to have a pure, clean, sweet flavor without any aftertaste. It has a low negative heat of solution and so does not produce the mouth-cooling effect of many other sugar alcohols, and it is practically nonhygroscopic. Isomalt has about half the sweetness of sucrose, and contributes about 2.0 kcal per gram. Reports indicate it is non-cariogenic. Diabetics easily tolerate isomalt because it results in insignificant changes in serum glucose and insulin levels.

Water solubility of isomalt is fairly low, about 25% at 20°C, and it is less hygroscopic than most other commercial polyols. In baking applications, it has been recommended that isomalt be substituted for sucrose on a one-for-one basis. Reduction in browning, as of crust, may occur.

A GRAS petition was filed in 1990.

Lactitol — Lactitol is a disaccharide sugar alcohol made by catalytic hydrogenation of milk sugar (lactose). Crystal forms include the anhydrate, a monohydrate, and a dihydrate; it is not very hygroscopic. It is currently being marketed under the trade name "Lacty" by a manufacturer in Holland. The substance is said to have a clean taste without aftertaste and its sweetness is about 40% that of sucrose.

Lactitol is not hydrolyzed or absorbed in the small intestine, and in the large intestine it is fermented into biomass and short chain fatty acids by the microorganisms present in that part of digestive tract. The net effect is that only about 2 calories per gram of lactitol is available to the body. Its consumption does not increase blood glucose or insulin levels, and so is well tolerated by diabetics (Blankers 1995).

It is also claimed that lactitol is noncariogenic, since oral microflora cannot use it as an energy source. As an alcohol, it is nonreducing and therefore would not be expected to participate in Maillard browning reactions.

The substance has been suggested for use as a bulk sweetener of relatively low calorie content in confections, chewing gum, frozen desserts,

Regulatory approval for use of lactitol in foods has been granted in a Japan, Australia, Canada, and a number of European countries. A petition for GRAS status was filed with the FDA in 1993. The scope of the original petition included use in chewing gum, hard and soft candies, frozen dairy desserts. and baked goods.

Maltitol — The polyol α-(1-4)-glucosylsorbitol is produced by enzyme hydrolysis of starch to give a maltose syrup, followed by hydrogenation of the maltose. Ingredients available commercially contain from about 88.5% to 99% of the compound, and are about 80% as sweet as sucrose. Water solubility, heat of solution, and effects on water activity and freezing point are comparable to these effects of sucrose. According to Oku (1994), the caloric content of maltitol, for humans, should be set at about 2 kcal/g.

Maltitol that reaches the large intestine is metabolized by bacteria commonly present to yield carbon dioxide, short chain fatty acids (which are utilized as energy), hydrogen, and methane. Secretion of insulin is only slightly stimulated by this sugar alcohol. Moderate intakes of maltitol, say over 0.3 grams per kilogram of body weight, have a laxative effect.

Maltitol has been shown to be more hygroscopic than sucrose, and therefore, less suitable for use as an ingredient in hard candy, since products containing maltitol become sticky and soften earlier or in drier atmospheres than candies made with only sucrose and maltose. An invention said to overcome this disadvantage was described in a patent obtained by Hirao et al. (1990). It describes a process to make anhydrous crystals of maltitol and a crystalline hydrogenated starch hydrolysis mixture solid containing anhydrous crystals of maltitol.

The firm Roquette American manufactures a maltitol syrup that can be substituted for corn syrup with a consequent moderate reduction in calories (about 25% as compared to an equal weight of sugar) and some reduction in sweetness. The trade name is "Lycasin." The same company distributes a crystalline version called "Maltisorb." The crystalline material has been used in Japan since the 1960s. A GRAS petition was filed in with the FDA in 1986,

A formula and process for making cream filling for sugar wafers is given in the Hirao et al. patent, as follows. A creamy product was prepared by mechanical blending of a mixture consisting of 2,000 parts of the novel crystalline material, 1000 parts of shortening, one part of lecithin, one part of lemon oil and one part of vanilla "oil." The material was held at 40°C to 45°C and sandwiched between wafers to obtain the desired product.

It has been shown to be possible to make bitter (semi-sweet?) chocolate using an anhydrous crystalline maltitol as a complete replacement for the sugar normally used. The composition of the material was 42% chocolate liquor, 13.5% cocoa butter, 44% maltitol, 0.48% lecithin, and 0.02% vanillin. The test materials processed satisfactorily through the critical refining and conching steps, and taste and texture properties of the finished product were very acceptable. Rapaille et al. (1995), who conducted these studies, estimated that calorie reductions of up to 23% could be obtained in commercial chocolates.

Maltitol has been used in combination with powdered cellulose (Solka-Floc 40 FCC) to enable the production of sugar-free caramel. The published formula and procedure is given below.

Premix
>   27.00% water
>   7.50% fat, 92°F melting point
>   0.25% lecithin
>   9.00% sodium caseinate
>   1.00% powdered cellulose
>   0.25% sodium bicarbonate
>   55.00% maltitol syrup, Hystar 5875

Final formula
>   23.37% premix from above
>   2.05% powdered cellulose
>   15.58% fat, 92°F melting point
>   57.35% maltitol syrup
>   0.50% salt
>   0.65% caramel flavor
>   0.50% vanilla flavor

Process

Premix: Mix the fat, lecithin, and water with vigorous agitation at 165°F. Dry mix the sodium caseinate, Solka-Floc, and sodium bicarboate. Heat the maltitol syrup to 165°F, then add the fat-water mixture followed by the dry mix. Mix vigorously in a vertical mixer for five minutes, then continue mixing in a blender for an addtional ten minutes. Refrigerate the premix overnight before using.

Final: Heat the maltitol syrup and salt to 280°F. Blend in the premix, fat, and Solka-Floc. Reheat the mixture to 260°F while stirring. Add the caramel and vanilla flavors. Pour onto slab, allow to cool, then cut into pieces of desired size.

Mannitol — This sugar alcohol is a white crystalline hexitol, occurring in three different optical configurations. Ordinary mannitol which, though levorotatory, is called d-mannitol, is found in nature as a constituent of the manna ash, celery, sugar cane, seaweeds of the genus *Laminaria*, etc. It is produced commercially by the catalytic hydrogenation of fructose. It is about 60% as sweet as sucrose, and has a caloric content of about 1.6 per gram. Mannitol has a negative heat of solution, as do most of the sugar alcohols, and so may give a cooling sensation when taken into the mouth.

Sorbitol — This hexahydoxy alcohol is found in nature as a minor constituent of the juices of several fruits, etc. It is produced commercially by the hydrogenation of glucose. It is about 60% as sweet as sucrose, and has a caloric density of about 2.6 kcal per gram. D-sorbitol has been used for a number of years in foods in the U.S., and currently is GRAS.

Xylitol — This five-carbon sugar alcohol is produced by catalytic hydrogenation of D-xylose, the raw material being produced from the xylan found in the hemicellulose fraction of birch trees. Xylitol is found in nature as a constituent of some fruits and vegetables; it is also a normal product of human metabolism.

Xylitol is about as sweet as sucrose, and has a pronounced cooling effect due to a relatively large negative heat of solution. The caloric density is said to be 2.5 kcal per gram, and it is evidently cariostatic and insulin independent. The sweetener is widely used as an ingredient in mints and lozenges produced in Scandinavian countries; in the U.S., it has been approved as a food additive for special dietary use.

Poorly Digestible Oligosaccharides

Among the useful ingredients for supplying volume and/or weight while reducing calorie content are several which we will call "Poorly digestible oligosaccharides." These are mostly polymers of glucose or fructose that , by their molecular structure or gross physical characteristics are not subject to enzymatic attack in the alimentary tract. Some of them can be used as nutrients by the bacteria normally present in the lower intestine, and part of the metabolic byproducts of these microorganisms can be taken up by the intestinal wall and used as energy or building blocks by the body. This means that some of the ingredients do, in fact, provide some calories to the system.

Alternan — Alternan is an unusual polysaccharide that has been recommended as a bulking agent or extender. It is said to have good water solubility and gives solutions of low viscosity. The substance is produced by selected strains of the bacterium *Leuconostoc mesenteroides*. Since these organisms ordinarily produce about equal amounts of dextran (arelated polysaccharide), purification of alternan is difficult and this has impeded its commercial production. Recent isolation and genetic improvement of a variant strain that does not produce dextran suggests that commercial exploitation of alternan is a possibility. These developments have been disclosed in a patent application of Leathers et al. (1995).

Alternan is resistant to enzymic attack, rendering it essentially non-caloric in foods. Its rheological properties are said to be similar to those of gum arabic, maltodextrins, and polydextrose. FDA status is not clear; there are some indications that the developers may claim GRAS status.

Inulin — Inulin is a nearly tasteless, white, semi-crystalline polysaccharide, composed primarily of fructose residues, that can be isolated from the sap of the roots and rhizomes of many plants. It resembles starch in many of its properties, and has been used for some decades in partly purified forms as a non-digestible analogue of starch. It can often be found on the shelves of health-food stores for those people who wanted to nourish their bifidobacteria. Because of its lack of utilizable calories, and its natural origin, it has attracted considerable attention as a bulking ingredient for dietary foods.

The dried, powdered tuber of the Jerusalem artichoke (from a perennial American sunflower *Helianthus tuberosus*) is being offered as a low-calorie bulking agent. It contributes only about 1.25 calories per gram, as compared to about 4 calories per gram from starch. The tubers contain about 65% carbohydrate, of which 78% is oligosaccharides of fructose and no starch, and 8% to 10% protein. The flour is slightly sweet. There is less than 1% fat in the material, on average. It is said that this dry powder can be substituted for up to 10% of the wheat flour in pasta and baked goods. In addition to slightly lowering the total calorie content, it is said to promote the growth of bifidobacteria in the colon, which is a great thing, as we all know (LaBell 1996).

Another form of oligofructose is the tradenamed Raftiloser, which is made by a Belgian firm by enzymatic hydrolysis of inulin extracted from chicory root. It is described as being a "fraction of inulin" and can be used in a liquid or powder format for carbohydrate replacement, or more specificially, as bifidogenic soluble dietary fiber. The material has a sweet taste with a "taste profile similar to sugar" and behaves functionally much like sugar syrups. The manufacturer says, "In addition to stimulating growth of bifidus bacteria in the intestinal tract, moderate amounts of oligofructose have been shown to have a beneficial effect on the HDL/LDL cholesterol ratio in blood serum, while reducing the serum triglyceride levels."

A Japanese company, Calpis, is promoting an oligosaccharide mixture derived from soybean "whey," the liquid byproduct of tofu production. The principal constituents of the oligosaccharides mixture are apparently raffinose and stachyose. The ingredient is about 70% as sweet as sucrose and contributes around 2 calories per gram. A characteristic thought by the developer to be important is that a daily intake of three grams of the material will greatly increase the number of Bifidobacteria in the intestine, and thus improve the general health of the consumer.

Polydextrose — In 1995, the Pfizer Food Science Group completed a factory for the manufacture of "Litesse," a trademark for their version of polydextrose. It is being promoted as a one-calorie per gram bulking agent for no-fat, low-fat, and reduced calorie foods. Polydextrose is a randomly bonded condensation polymer of dextrose that was developed specifically as a low-calorie bulking agent (Deis 1993). Early versions had a somewhat sour taste, due evidently to impurities; the present versions have a much lower titratable acidity.

Polydextrose is readily soluble in water, so that solutions of about 80% concentrations can be obtained at 25°C. It is reported to be extremely stable to most processing conditions, and to be noncariogenic. It was approved as a food additive in 1981, and is permitted in baked goods and baking mixes, confections and frostings, gelatin desserts, puddings,etc. In the U.S. Litesse is used in low fat ice creams, yoghurts, confections, and some other food products. Keebler is using the ingredient in a fat-free line of fruit bar cookies.

Table 3.1
REPLACING NUTRITIVE SWEETENERS IN HARD CANDY[1]

Ingredient	Control formula	With PD	With PD & isomalt
Sucrose	59.00	38.82	--
Corn syrup	18.64	--	--
Sodium citrate	0.10	0.10	0.10
Water	20.51	20.51	20.51
Citric acid	1.50	1.50	1.50
Flavor	0.15	0.15	0.15
Color	0.10	0.10	0.10
Polydextrose	--	38.82	38.82
Isomalt	--	--	38.82
Energy content	288	194	117

[1] Quantities are in percent of the batch weight (except for calories), on as is bases. Calories are total kcal per 100 g. PD = polydextrose.
Source: Kosmark, as quoted by Giese (1993).

In hard candy, where sweeteners (usually predominantly sucrose) provide the basic structure of the product, selection of a bulking agent with satisfactory functional properties but having much reduced caloric content is restricted to only a few of those listed in this section. It appears that use of polydextrose, or polydextrose in combination with isomalt, leads to acceptable products. The formulas in Table 3.1 are attributed to Kosmark (quoted in Giese 1993).

Resistant Starch

The term, "resistant starch" has come into prominence in recent years, partly as a side issue in the race to maximize the content of dietary fiber in food. It is generally understood that resistant starch is a substance that has the chemical structure of normal starch (amylose or amylopectin) but, due to a difference in its physical structure, is not digested, or is digested incompletely in the human alimentary tract.

In addition to "resistant starch" we also hear of "slowly digestible starch," which is nothing more than a starch that falls somewhere between resistant starch and readily digestible starch. No doubt there is a continuity between the extremes in which would be found innumerable variations.

What causes starch to be resistant to digestion? It is known that the physical state of starch may vary in accordance with the genetics of the plant producing the material, the processing treatments applied to the starch during food preparation, and retrogradation of gelatinized starch during storage or dehydration of solutions. In fact, however, many aspects of the formation and digestive fate of resistant starch are still unknown.

Natural starch — Starch occurs in plants and some microorganisms as partially crystalline granules composed largely of two polymers of glucose: amylose and amylopectin. Amylose is a mixture of linear and slightly branched polymers, with the latter having 5 to 20 chains attached to the main string. Amylopectin is a highly branched, high molecular weight polymer with a structure composed of clusters with chains of approximately 14 to 16 glucose units in length associated in double helices; about 80% to 90% of the chains are localized in single clusters, with the remainder forming connections between the clusters. The clusters in amylopectin form the crystalline regions of starch while the intercrystalline regions containing branch points form the amorphous portions. Amylose and amylopectin are intermingled in the starch granule with a tendency to show a radial orientation from the hilum of the granule. Gelatinization of starch, the result of heating it in contact with water, involves a transition from the ordered arrangement in the granule with as the crystallites unfold.

Molecules having the structure of starch, as determined by traditional methods of analysis, can exist as semi-crystalline native granules, as partially disordered molecules after gelatinatization, or as partially crystalline retrograded molecules. The physical differences, largely the result of the type and extent of intermolecular aggregation, can lead to variations in starch's response to enzymes.

Digestibility of starch — Salivary gland enzymes and enzymes secreted in the small intestine break normal starch molecules into dextrins. These relatively small compounds are hydrolyzed to give glucose by cells located in the intestinal lining, and the glucose molecules are transferred into the blood stream. Resistant starch does not respond normally to either the salivary enzymes or the intestinal secretions. Consequently, some or all of the material may reach the large intestine more or less intact. There it can be attacked and broken down by enzymes of the microflora to give fermentation products such as volatile fatty acids, which reduce the pH. Starch not utilized by the microorganism may contribute to fecal bulk, and can thus be considered "dietary fiber."

Several different starch structures can be found in processed foods. As a result of certain kinds of thermal treatment or during prolonged storage of gelatinized starches, a number of physical changes occur that are called retrogradation. This is a sort of quasi-crystallization process that affects only parts of the large starch molecule; the amylose chains (or some of them) interconnect forming junction zones including double helices, and the amylopectin chains rewind into double helices. In pure starch gels it is accompanied by the development of increased opacity and shrinkage of the gel with extrusion of fluid. It can be detected by X-ray diffraction studies. The basic mechanism is the result forming of hydrogen bonds between hydroxy groups on adjacent starch molecules.

The gross changes resulting from extensive retrogradation have major impacts on the texture and digestibility of foods. For example, the firming of bread that is called texture staling is apparently due in large part to retrogradation.

Although definitive data are lacking, it has been hypothesized that consuming increased amounts of resistant starch may decrease the chances for suffering colon cancer (Muir et al. 1996). The reasoning goes something like this: (1) International surveys have found that a strong inverse correlation exists between starch consumption and the incidence of colon cancer; (2) The protective effect may be related to a fraction of the starch consumed in a high starch diet that reaches the colon undigested. In the cited study, human volunteers were fed two experimental diets, the principal variable being the levels of resistant starch, in one case about 5 grams per day, in the other about 39 grams per day. The high RS diet produced a 43% increase in fecal output and lowered the fecal pH by 0.6 units. There were also significant increases in the fecal concentration and daily excretion of the short chain fatty acids. The high RS diet also reduced fecal concentrations of "potentially harmful" phenols and ammonia.

Resistant starch as an ingredient — It has been shown that controlled physical modification of starch can result in a product that is resistant to the enzymatic processes occurring in the small intestine. Various methods have been suggested for converting corn starch, for example, into a product having many of the desirable functional properties of the native material, but being much less available to the digestive processes, i.e., low calorie or no calorie. One such method was revealed in the patent of Iyengar, et al. (1991), describing a food-grade, non-digestible low-calorie bulking agent derived from starch. The process involves retrogradation of starch, followed by enzymatic or chemical hydrolysis to reduce or remove the amorphous regions of the starch molecules. The modified starch that results can be regarded as a retrograded starch containing few or no amorphous regions. The properties of the product, including its caloric value and water-holding capacity are said to be controlled by varying the degree of modification at either step, i.e., the degree of retrogradation and/or amount of hydrolysis.

The correct label nomenclature for resistant starch, according to current advice, is "maltodextrin." As a form of maltodextrin, it is presumably GRAS.

Table 3.2
FORMULAS FOR HARD SUGAR COOKIES
WITH AND WITHOUT RESISTANT STARCH

Ingredient	Control	Experimental
Flour	40.50	13.12
Shortening	19.30	25.03
Brown sugar	14.40	18.67
Granulated sugar	14.40	18.67
Whole egg solids	2.30	3.00
Water	7.40	9.60
Salt	0.60	0.78
Baking soda	0.60	0.78
Vanilla extract	0.50	0.65
EMRA	---	9.70

EMRA = crystalline water-insoluble enzyme-modified retrograded amylose of the invention of Iyengar et al.
The shortening was Creamtex partially hydrogenated vegetable oil.
Source: Iyengar et al. (1991).

Preparation procedure for Table 3.2:
1. Hydrate the powdered eggs in water (2.3 g egg to 7.4 g water).
2. Cream the salt, sugar, baking soda, and shortening.
3. Add the hydrated eggs slowly while mixing.
4. Add the remaining dry ingredients and mix for about 30 seconds.
5. Add the vanilla and the remaining water and mix well.
6. Place 15 g portions on an ungreased baking sheet and bake at 375°F for about 8 minutes.
Results: The test cookie was hard with a highly acceptable texture.

Bakery products made with resistant starch — In their patent, Iyengar and co-workers (1991) disclose the results of a number of tests in which bakery products (cookies) were made with resistant starch replacing much of the flour. Unfortunately, a numerical value for caloric reduction was not given in the patent, but they indicate that some of their resistant starch ingredients were only 7.8% digested by pancreatin by the time 100% of a soluble starch was digested.

Table 3.3
FORMULAS FOR REDUCED-FLOUR BROWNIES
WITH AND WITHOUT RESISTANT STARCH

Ingredient	Control	Experimental
Shortening	16.00	16.00
Granulated sugar	36.47	36.47
Eggs	16.75	16.75
Unsweetened chocolate	10.55	10.55
Baking powder	0.24	0.24
Salt	0.57	0.57
Vanilla extract	1.00	1.00
All-purpose flour	18.42	9.41
EMRS	---	9.41

EMRS = high amylose starch retrograded and enzymatically modified according to the invention of Iyengar et al.
Source: Iyengar (1991).

Preparation procedure for Table 3.3: 1. Melt the shortening and chocolate together at low heat with constant stirring. Allow to cool. 2. Beat eggs until light, then blend into the sugar. 3. Slowly beat the cooled chocolate-shortening mix and vanilla into the sugar/egg mixture. 4. Slowly blend all of the dry ingredients into the previous mixture. 5. Bake for 35 minutes.

Results: Brownies containing the resistant starch exhibited a slighly chewier and drier texture, compared to control, but were generally acceptable.

Table 3.4
FORMULAS FOR REDUCED-FLOUR SUGAR COOKIES
WITH AND WITHOUT RESISTANT STARCH

Ingredient	Control	Experimental
Shortening	21.00	21.00
Granulated sugar	23.92	23.92
Fresh egg	7.32	7.32
Whole pasteurized milk	3.47	3.47
Baking powder	0.65	0.65
Salt	0.24	0.24
Vanilla extract	0.44	0.44
All-purpose flour	42.96	24.48
EMRS	---	24.48

EMRS = high amylose starch retrograded and enzymatically modified according to the invention of Iyengar et al.
Source: Iyengar (1991).

Preparation procedure: 1. Cream together the shortening, sugar, and vanilla. 2. Add egg to creamed mixture and beat until the batch is light and fluffy. 3. Blend all the dry ingredients together, then sift and mix them into the creamed mixture. 4. Chill the finished dough for one hour before baking. 5. Sheet out the dough, cut shapes, and place on a greased cookie sheet. 6. Bake at 375°F for 6 to 8 minutes.

Results: The test samples had a slightly chewier texture than the control, but the product was acceptable and generally comparable to the control.

Commercial availability of resistant starch — In about 1994, suppliers began to offer resistant starch in commercial quantities for use in dietetic foods. These items appear to be a form of retrograded amylose. The brand name Novelose has been used for this product. It is currently available from National Starch and Chemical Co. and Opta Food Ingredients Co. Kevin (1995) says the label declaration can identify these ingredients as "maltodextrin." The dietary fiber equivalent is about 30% of the weight and they yield 2.8 cal/gram for labeling purposes. They have small and uniform particle size and limited water-binding capacity. They have been recommended for use in extruded snacks, RTE breakfast cereals, cookies and crackers.

chapter four

REDUCING FAT CONTENT: GENERAL CONSIDERATIONS

Fats are very common food ingredients, and have been highly regarded as improvers of taste and texture and as valuable sources of energy throughout historical times. Many different types of culinary fat have been used, the earliest probably being the drippings of roasted meat. Great fortunes were made in the ancient world in pressing olive oil and transporting it to wherever a demand existed among people who could pay for it. Nowadays, we have many sources: oils from palm, coconut, soybean, rapeseed, cottonseed, sunflower seed, safflower seed, olive fruits, and corn germ; fats from cattle, hogs, sheep, fowl, marine mammals, fish, etc. Each of these materials has distinctive physical, chemical, and nutritional properties that must be taken into account when selecting an ingredient for a specific application.

In a chemical sense, fats and oils are triglycerides of fatty acids. When oils are mentioned in this book, lipids from animal or vegetable sources are meant. Mineral oils are very different from food oils, being mainly hydrocarbons which are not utilized for any nutritional purpose by the human body; they are of no significance as food ingredients and will not be discussed in detail in this book.

Fats extracted from either vegetable or animal sources will contain small amounts of materials such as mono- and di-glycerides, free fatty acids, tocopherols, sterols, phosphatides, and fat-soluble vitamins. Other impurities contributing color and flavor may also be present. The predominant sterol in animal fats is cholesterol, which has been implicated in human atherosclerosis and other disorders, and which will be discussed in considerable detail in a subsequent chapter. It is generally believed that vegetable fats never contain cholesterol though it is well-established that they contain other sterols.

Extraction, purification, refining, molecular rearrangement and other operations that are applied to natural fats in order to convert them into the shortenings, frying fats, and other ingredients of commerce, remove most if not all of their impurities. Processing may also deliberately or unintentionally cause chemical changes in the fatty acid portions of the fat. In particular, hydrogenation (a very common procedure) causes all or part of the unsaturated fatty acids to become fully saturated, a significant change affecting their nutritional qualities as well as their suitability for various culinary purposes.

REDUCING FAT IN THE DIET

Since it is quite obvious that fats and oils, with their typical nine calories of energy content per gram, create major problems in the designing of diets for persons desiring to reduce body weight, these foods and ingredients are among the most favored candidates for elimination or change by nutritionists seeking to encourage more helpful eating habits among the American public. An unfortunate impediment to these attempts is that foods in which the expected or customary amount of fat has been greatly reduced, and the meals in which they are included, often have very low acceptability.

There can be little doubt that those industrial food technologists trying to find niches for new products of the dietetic type have directed much of their resources toward reducing the fat content, and therefore the calorie load, of foods that will be consumed by average healthy consumers. That is, modification of the type of fat for treatment of specific diseases or prevention of various disorders other than obesity has played a rather minor role in dietetic product development.

This is an understandable bias, since the number of persons desiring to reduce weight, or to retain a desirable weight, by dietary means far exceeds the number of persons known to suffer malfunctions of their lipid metabolism. In addition, it may be considered that manipulating lipid species for the purpose of alleviating disease comes dangerously close to the medical/pharmaceutical area with the possibility of encountering all the regulatory delay and enormous testing expenses that are involved in the introduction of a new drug.

Adjusting Lipid Content or Type

The simplest approach to reducing calories from fat content is to cut back on the percentage of fat in an original and commercially successful formula, while retaining the relative proportions of the other (nonfat) ingredients. This is difficult to put into practice because the fat content plays many roles, both in processing and in consumer response. If the traditional product is sprayed with fat after baking (as with some crackers) or fried, the task is somewhat simpler, but even in such cases, appearance, texture, and some other factors affecting acceptability are related to fat content. It can be expected that any substantial reduction in the fat uptake during frying or the fat applied by spraying a baked product will cause some customer dissatisfaction. After all, the fat content in either case has presumably been optimized over a period of years by successful marketers and any reduction in the current amount will probably decrease consumer satisfaction.

As an example of the problems faced by formulators, let us consider the case of a technician assigned a project to reduce the fat content of a well-accepted brand of cookie. The elasticity of gluten, from flour, is objectionable in those bakery products where a tender, "short" texture is expected. This undesirable feature is traditionally offset in products such as cakes, cookies, pie crusts, muffins, and the like by adding relatively large amounts of shortening, i.e., either plastic or liquid fats. If the fat content is lowered to decrease the caloric content of the cookie, other means of tenderizing, softening,and otherwise optimizing the finished product must be used. Major changes in composition may be required. Seyam (1996) formulated fat-free or low-fat cookies with decreased rubberiness (due to gluten) by replacing at least 50% of the wheat flour with a mixture of at least 30% by weight of white rye flour, at least 30% by weight of corn flour, and 10% by weight of rice flour.

If the principle is carried to an extreme by baking products that are normally fried (potato chips, corn chips, doughnuts), substantial decreases in both calorie content and fat content can be achieved. Such products have been marketed, and they appear to have achieved some success, which may, however, be limited demographically and in duration.

Replacing Fat Functions with Lower Calorie Substances

In many food products, the contribution of fat to the total caloric load of the finished material considerably exceeds the contribution made by both carbohydrates and protein. In the baking industry this is particularly true in the case of pie crusts, many cookies, most cakes, and virtually all fried products. On the other hand, a few items, such as angel food cakes, contain no significant amount of fat. Even if the fat must be replaced gram for gram by carbohydrates or proteins, a significant savings in calories might be achieved in selected products.

This situation has provided the impulse for much research into means for replacing ingredients of the fat and oil type with low calorie or no calorie substitutes. There have been two major approaches: (1) the fat is replaced by hydrocolloids or similar materials that increase aqueous phase viscosity so as to imitate the mouthfeel and/or viscosity of oil — a strategy that has been very successful in formulating reduced-calorie salad dressings, and perhaps could be useful in formulating reduced calorie icings, toppings, etc., and (2) fats are replaced by a molecular species having some of the same physical characteristics as fat but having little or no caloric value when ingested by humans.

Strategies and principles — Some consumers, perhaps a majority, have been conditioned to believe that the most healthful food is the product with the least amount of fat. Many ingredient suppliers and bakery formulators, recognizing that this demand exists, have spent a great deal of effort in developing materials and devising formulas that have the sole purpose of reducing the fat ingredients of popular foods to essentially zero. A considerable amount of success has been claimed, and some has actually been achieved, in these studies. As a result, dieters have a myriad of fat-free or reduced-fat bakery products to choose from — this no doubt reduces their caloric intake considerably, especially if they find the products to be inedible after they start to consume them, as they undoubtedly will in many cases. It is clear enough that, in many cases, much better sensory properties could be achieved in the finished products if only one or two percent of fat were included, accepting the small increase in calorie content this causes.

A set of "Rules of Thumb" was suggested by Bakal (1994) for designing reduced fat products. These are:

• Increase sugar content and emulsifiers in baked goods. Use brown sugar or fructose, together with starches and gums, to increase humectancy.

• Use combinations of fat mimetics. Effective pairings include rice starch and gelatin in spreads; starch and cellulose gel in spoonable or pourable dressings; and cellulose gel or other gums with protein-based mimetics. Consider rice starch as an opacifier. Polydextrose should also be considered, especially if calorie reduction is an objective.

• Add a small amount of fat. Levels as low as 1% have significant impact on flavor, mouthfeel, and consumer acceptance.

Varieties of bread having low contents of fat can be formulated readily; problems are minor because ordinary white loaf bread is not very high in fat content. Leaving out the shortening saves a few calories per serving, but the reduction in fat is probably of more interest to persons who are specifically reducing their intake of triglycerides to the lowest possible level than it is to weight-reducers. Both the processing response and the eating quality of the bread will be affected by fat reduction, but not necessarily to the extent that either factor falls into the unacceptable range. Use of emulsifiers or surfactants in less than 1% quantities can improve dough and bread quality in low-fat products, and these ingredients do not necessarily increase the triglyceride content of the bread. However, diluents, carriers, and impurities found in commercial versions of emulsifiers may be triglycerides.

The nomenclature used in this chapter and the next two chapters may vary somewhat from that used in the popular press. We have defined "fat mimetics" or "fake fats" to mean synthesized chemical compounds that have been designed to resemble triglycerides of fatty acids. The term, "engineered fats," refers to triglycerides of fatty acids that, because of chain length or other features, have properties important to the formulator of dietetic products. A category we call "fat substitutes" or "fat replacers," includes the many ingredients relying on hydrocolloids, emulsifiers, etc., that permit the addition of relatively large amounts of water to provide texturizing effects that duplicate at least some of the sensory impressions normally produced by ingredient fats and oils.

Food Types that Could Benefit from Fat Reduction

The Calorie Control Council of Atlanta GA published a table listing the types of fat replacers that could be used in various categories of foods (Anon. 1996A). This compilation has been modified and rearranged to give Table 4.1. In some cases the fat replacer may be the only ingredient that has to be changed to achieve the desired result, while, in other cases a combination of ingredients may be needed (e.g., bulking agents) to complete the formula.

Fat Substitutes and Replacements

When a low-fat product cannot be formulated simply by reducing the amount of oil- or fat-containing ingredients in a portion of a finished food, because the taste, texture, appearance, shelf-life or processing response of the reformulated product is unsatisfactory, there are two main approaches that have been used to design substitutes: (1) Use a fat-mimetic, or fake fat, that has many of the physical and chemical properties of normal fats and oils but makes little or no caloric contribution to the diet, and (2) Use a non-lipid ingredient that imitates the textural contribution of fat (greater lubricity, increased viscosity, etc.) by binding water. Akin to the first approach is the use of short-chain triglycerides, which are genuine fats but generally not natural fats, to replace the usual long-chain triglycerides found in the more common types of fats and oils. This strategy does save some calories, but only a relatively small percentage on a weight-for-weight replacement basis. Some writers have called these triglycerides "engineered fats."

To some extent, emulsifiers such as monoglycerides may compensate for lowered fat content by making the lipids which are present in reduced amounts in the finished food more effective in modifying and controlling its texture; these materials can be considered as straddling the two categories, but are more closely connected with the first. In addition to the many kinds of emulsifiers used in foods, there is a somewhat similar group of synthesized wetting agents. Some emulsifiers are closely allied chemically to lipids, while others are far different in chemical composition.

In the category of fat mimetics, we find such materials as Olestra that bear some chemical and physical resemblance to fats but are not triglycerides of fatty acids and are essentially indigestible, and in the second category we have a large number of hydrocolloids (gums, starch, etc.) that form viscous solutions with water and may or may not contribute calories.

Table 4.1
TYPES OF FAT REPLACERS FOR DIFFERENT CLASSES OF FOOD

Carbohydrate-based	Protein-based	Lipid-based
Dairy products: beverages, cheese, sour cream, etc.		
Celluloses & gums	Microparticulated	Emulsifiers
Maltodextrins	protein	
Starches	Other protein-	
Polydextrose	based ingedients	
Refrigerated & frozen desserts: ice cream, puddings, cheesecake		
Cellulose & gums	Microparticulated	Emulsifiers
Maltodextrins	protein	Lipid analogues
Starches	Protein blends	
Polydextrose	Other protein-	
	based ingredients	
Soups, sauces, gravies		
Cellulose & gums	Microparticulated	Emulsifiers
Maltodextrins	protein	Lipid analogues
Starches		
Margarine, pourable dressings, mayonnaise, and other spreads		
Cellulose & gums	Microparticulated	
Gelatin	protein	Emulsifiers
Maltodextrins, starch		Lipid analogues
Polydextrose		
Bakery products		
Cellulose & gums	Microparticulated	Emulsifiers
Maltodextrins	protein	Lipid analogues
Starches		
Polydextrose		
Confectionery and candies		
Cellulose & gums		Emulsifers
Crystalline fructose		Lipid analogues
Maltodextrins		Caprenin
Starches		
Polydextrose		

Also in the latter class, are the fruit pastes and concentrates containing pectic and cellulosic substances that, in combination with water, give some of the organoleptic properties normally resulting from the use of shortening. They may contribute significant amounts of calories from carbohydrates and protein.

Another approach, somewhat similar to fat mimetics, but having important differences is illustrated by a product having the trademark name "Simplesse." This material is made by a patented microparticulation process of heating and blending whey proteins (egg white protein can also be used). It consists of large quantities of uniform spheroidal protein particles ranging from 0.1 to 3.0 microns in diameter. These globules do not coalesce under ordinary circumstances and they are stable under a fairly wide range of conditions. They can be heated and used over a pH spread of about 3.5 to 7.0 without disrupting the particles. They retain water, diffract light, interrupt food matrices, and exhibit viscoelasticity. They are said to interact with buccal cavity surfaces to impart the perception of creaminess associated with emulsified fat systems. Simplesse is available as a dry powder.

When used as an ingredient to replace fat, Simplesse has a viscous, lubricating type of texture and appears to function very well in certain specialized applications — as ingredients in no-fat ice cream, as cream replacers, etc. Simplesse probably would be of little or no use in a baked or fried dough, but might be useful in adjuncts such as buttercream icings, whipped toppings, salad dressings, etc. These materials not be expected to be of value as a frying medium.

Simplesse has had to surmount many legal obstacles, and after many years its manufacturer secured regulatory approval to use the ingredient in several commercial products. It does not seem to be a major factor in the low calorie ingredient market at the present time, possibly because of its relatively high cost compared to some of the other fat replacers.

What about flavor? — Until they become rancid, normal fats do not have much odor or taste. Some authorities have said that completely purified lipids have no odor, while others say the odors of highly purified oils may be faint but can be detected. Since most food technologists will not be working with highly purified ingredients, the organoleptic characteristics of such materials become an academic question.

The flavor of fats is generally fairly easy to duplicate by additions of concentrated natural or artificial substances. The situation with butter is typical; here the principal flavoring materials are held in the microscopic droplets that are dispersed throughout the continuous fatty phase, and they can be duplicated and concentrated by fairly simple methods. Very good natural and synthetic butter flavors are available at relatively low prices from many flavor supply companies.

If sufficient demand develops, similar flavoring materials could be developed and become commercially available to duplicate the character-istics of other oils such as olive oil, etc. A faint tallowy (hydrolytic) rancidity seems to be a normal component of the flavor of some ingredients, such as coconut; these notes, too, can be duplicated in concentrated form without much difficulty by capable flavor compounders. At any rate, reproducing the flavor contribution of the fatty component is one of the lesser problems facing the formulator of low-fat foods.

Perhaps a greater problem is the retention of fat-soluble added flavors during processing and storage. Intermediate products and finished products that contain only traces of lipids will quickly lose many fat-soluble flavor compounds that are normally trapped in emulsion globules and hindered from evaporating or reacting with oxygen or other contacting materials. Many, though not all, added flavor molecules have lipophilic tendencies and some of the Maillard reaction products formed in the crust during the latter stages of baking are definitely lipophilic.

Fake Fats

The group of fat replacers sometimes called "fat mimetics" or "fake fats" includes, but is not limited to, sucrose polyesters, polyglycerol esters, triglyceride esters of alpha-substituted carboxylic acids, and alkyl glycoside polyesters. These are not triglycerides of fatty acids, and so can be expected to differ in numerous respects from true fats. Many of them furnish zero calories when ingested.

Probably the most work in this category of substances has been done with sucrose polyesters. Among these is "Olestra," the trade name for a mixture of sucrose molecules that have been reacted with six, seven, or eight long-chain fatty acids. Olestra "looks like fat, cooks like fat, and gives food the rich taste and mouth feel of ordinary fat," says one writer (Raber 1995). It is not digested in, or absorbed from, the human intestinal tract, so it contributes no calories to the diet.

Olestra, and possibly other sucrose polyesters, are differentiated from most fat substitutes by their fairly good stability at high temperatures, such as the heat encountered in baking or frying. Other possibilities for their use have been listed by Nelson (1990).

There appear to be some negatives to Olestra's use in foods; critics say the substance causes abdominal cramping and diarrhea, and the the liquid material leaks from the digestive tract. The substance also inhibits the absorption of some vitamins and other nutrients in the digestive tract. The effect of the relatively very stable Olestra on the environment as it accu-mulates in waste, and in particular its effects on water quality and wastewater treatment processes, has apparently not been examined in any detail (Giese 1996).

Reducing the Fat in Fried Products

Much of the caloric content of fried goods such as doughnuts is due to absorption of the fat that is used as a heat transfer medium. The content of triglycerides in the raw products, before frying fat has been absorbed, is often fairly low. Most of the fat is absorbed in the final stage of cooking, when the crust is turning brown. Prior to that time, the rapid evolution of water vapor and carbon dioxide from the hot dough tends to restrict contact of the frying medium with the dough surface so that opportunities for absorption are limited.

A relatively small reduction in the uptake of fat can be achieved by careful control of the frying conditions, and by modifications in the size, shape, formulation, and other features of the raw product. Some of these changes may adversely affect the acceptability of the cooked product and others are not practical from a production standpoint.

An obvious approach to calorie reduction would be to use a frying medium that contributes fewer calories than the usual type of fat, on a equal weight basis. This alternative has been difficult to achieve, however.

Changing the formulation or the cooking process – Breaded products such as chicken parts that are cooked in a vat of hot oil are high-fat foods in their finished form. Large amounts of frying oil are absorbed by the porous, crisp outer layer formed in the late stages of the cooking process. Many patents can be found that claim to greatly reduce the fat absorption encountered in normal frying operations.

A simple procedure that often reduces the fat content of fried foods by a few percent is to subject the cooked product to vigorous agitation and a strong current of air immediately upon the product's removal from the frying kettle. These steps remove adhering fat not yet absorbed into the product. All frying operations have a draining step, but few recognize the importance of using techniques necessary to remove all liquid fat before it has an opportunity to solidify or soak into the substrate.

Some methods for reducing fat content in fried foods rely on breadng mixtures that either brown very quickly or rapidly form a more or less impervious outer layer that hinders penetration of the frying medium. An approach to this method (Chalup and Sanderson 1996) involves mixing flour, water, dextrose, nonfat dried milk, sodium salt, calcium salt, and gellan gum to form a gellan gum batter. The food substrate is coated with the batter, then frozen, and cooked in a conventional manner.

French fries containing 45% less fat than the conventional product are said to be achievable using a pectin-based ingredient (Slendid 400) from Hercules Inc. Although the potatoes can be fried in a generally traditional manner, the coating reduces absorption of oil during the cooking process. The finished product is said to retain natural potato flavor, texture, and appearance. No changes in cooking procedures or equipment are required, according to the manufacturer of the ingredient.

The production of reduced-fat fried snacks with lighter, more expanded structures is disclosed in the patent of Villagran et al. (1995). The process requires forming a sheetable dough consisting of 1% to 10% calcium carbonate, 50% to 70% of a source of starch-based flour comprising pre-gelatinized starch, at least 3% hydrolyzed starches having dextrose equivalents of about 5 to 30, up to 5% emulsifier, and about 20% to 40% of added water. After the dough has been sheeted into the required thickness, snack pieces are cut from it. No doubt, the highly expanded piece described by the inventors absorbs less frying oil because the continuous evolution of gas until a late stage in the frying process keeps the frying medium from penetrating deeply into the structure. Generally, though, larger surface areas per unit weight lead to more absorption of frying fat.

Using a somewhat different approach, Kaslas et al. (1994) propose to fry the food in a more or less normal process, then extract most of the fat content by supercritical carbon dioxide extraction. The food is placed into a supercritical fluid extraction vessel with liquid carbon dioxide at a temperature of 0°C to 30°C. Pressure in a range between the carbon dioxide liquefaction point and 4,500 psig is applied to the container's contents. The inventors say that about 50% of the oil is removed from the food and enters the carbon dioxide, which is then removed from the vessel. This type of processing would seem to be prohibitively expensive for most food manufacturers.

A synthetic hydrocarbon-based cooking oil was developed by scientists working for Mobil Oil Corp. Low et al. (1994) patented this non-toxic hydrogenated oligomer of an alpha-olein having 5 to 20 carbons. It is evidently non-digestible and thus has zero calorie content so far as the human digestive system is concerned.

Eliminating the frying process — A more direct approach has been to eliminate frying altogether, cooking the snacks by dry heat methods such as oven baking, microwave heating, and cooking with radiant heat. There are several patents and publications describing the finishing of snacks and conventionally fried dough products by microwave heating, forced air drying, and resistance heating. These methods can certainly set up the interior structure and remove the desired amount of water, but they seldom provide a finished product that resembles the foods consumers have come to know and love. The major problem appears to be obtaining the color, texture, and flavor that is normally created in the surface layers of the product by the frying medium.

Considerable ingenuity has been shown in devising surface treatments to provide a the desirable crispness and a proper color. Flavor deficiencies are harder to solve. Not only has commercial success been limited, but production lines usually become more complex when switching to dry-cooked simulations of fried foods. Also, shape, dimensions, and density of the product may have to be altered from the traditional style in order to maximize the effects of the heating technique, and multiple heating-cooling stages are often required.

At least some non-fried replacements for snack chips and fried doughs have received moderate acceptance because their marketing approach emphasized the fat reduction and/or calorie reduction. Without this gimmick, however, it is doubtful that such products could be successful.

Fairly typical of the methods that have been used to replace frying, is the process described in the patent of Duncan (1990), which applies mostly to breaded meat items such as poultry, red meat, and fish. The food item is first wetted and then coated with a breading mixture. After spraying the food items with a mist of water until the breading composition appears sticky, the pieces are put on a cooking tray. In the cooker, hot gas [air?] jets are directed against the coated surface of the items. The preferred temperature range is 375°F to 475°F, and the cooking time is 11 to 25 minutes. After heat treatment has been completed, the pieces are sprayed with a mist of water to turn the item uniformly brown. The inventor says, "Food which has been cooked by this method can be vacuum packed and retains its flavor for an extended period of time."

Middleby Marshall, an old-line oven manufacturer, has introduced a hot air cooking device that is said to turn frozen potato strips into a product resembling french fried potatoes. The cooker senses the moisture content and temperature of the potatoes, which are tumbled inside a basket, and follows the instructions of a microprocessor to adjust the supply of 500°F air. In about 3 to 4 minutes, the "fries" are browned on the outside but soft on the inside. This appears to be a unit intended for use in home kitchens or in vending machines

The patent of Lewis and Lewis (1995), teaches a process for preparing low- or no-fat potato chips or straws by a method that does not include a frying step. As an initial step, raw potatoes are sliced to form 2-mm to 3-mm slices that will not break or become excessively hard on subsequent dehydration. The slices are blanched and washed to remove excess free starch, dehydrated in hot air until they reach a moisture content between 12% and 30%, then toasted at 140°C to 220°C until they reach a moisture content below 4%.

Extrusion puffing — Extrusion cooking with limited puffing can give varying degrees of crispness in finished products, and by this technology shapes can be made to match those of many kinds of fried chips. It is customary to spray these puffed pieces with a small amount of flavored oil subsequent to extrusion; the fat helps disguise the basically powdery texture and also gives a more rapid and more intense flavor impact when the product is consumed.

The eating texture of extruded foods without fat cannot match the acceptability of the usual commercial puffed snack product. The flavor also is seldom a close duplicate to that of, say, traditional potato chips or tortilla chips. In addition, the contrast between the rather flinty texture of the outer layer and the somewhat more fragile structure of the interior, which has a certain appeal (e.g., tortilla chips and potato chips), is difficult to achieve in extruded items.

The free oil in a fried snack has an effect on both flavor and texture. It acts as a distributor or carrier of flavored compounds as well as contributing some flavor of its own and interacting with the base material to form other flavored compounds. Spraying extrusion-puffed snacks with oil can partially overcome their deficiency in these respects, but the obvious need to keep the spray oil to a minimum may prevent a close duplication of fried goods. Another strategy is to spray on a fake fat in the hope or expectation that it will provide the needed textural and flavor characteristics without contributing calories.

All of these approaches have been tried at one time or another, and some of them have been fairly successful so far as customer acceptance is concerned.

ANALYTICAL METHODS FOR LIPIDS

Designing a food with reduced fat content or with specialized lipid profiles for dietetic purposes is inextricably connected not only with prior study of the relevant labeling regulations but also with an understanding of the way the food's composition will be determined by laboratories of the regulatory agencies that will be testing it. Details of analytical methods and, in particular, their specificity and precision, are definitely pertinent to a discussion of the legal status and thus of the marketability of any new product.

This section will give a brief survey of the current status of procedures for determining total fat and fatty acid profiles. This technology is in a state of flux and constant review of the literature of the field is highly desirable. For small companies, it is possible to resort to consulting laboratories which have the requisite equipment and personnel for applying the tests and which have developed skills in both lipid chemistry and regulatory affairs.

As a starting point, a brief summary of the types of tests normally applied in the quality control testing of food oils and fats may be helpful. The following section is based on the articles of Pomeranz and Meloan (1978), Potter (1986), Weiss (1970), Newton (1989), and Dziezak (1989), as modified, corrected, and expanded. The order in which the tests are arranged has no special significance.

Procedures for Quality Control and Characterization

Many of the tests applied to fats and oils, and particularly to raw materials such as oil taken from the presses or extraction vessels during the production cycle, are highly empirical in nature and the results vary according to the conditions of the tests. In recent years, analyses based more on instrumental determinations of the basic physical and chemical properties of the oils and fatty acids have come into use and have greatly improved the reliability and meaningfulness of results. Some of the characteristics traditionally regarded as giving a basis for evaluating the quality of fats and oils are listed below.

Acid value — measures the free fatty acid content and has been used as an index of hydrolytic rancidity development; involves the determination of the amount of potssium hydroxide required to neutralize one gram of fat or oil. Acid value is used to monitor refining operations.

AOM (active oxygen method) — gives an indication of the oxidative stability of a fat or oil; filtered air is blown through a heated and moisture-free sample, and peroxide values are determined periodically.

Bleaching test — measures residual color; used primarily for cottonseed, soybean, and sunflower seed oils.

Carbonyl test — determines carbonyl-type compounds developed at a primary stage of lipid oxidation.

Cloud point — determines the temperature at which a cloud forms in an oil sample immersed in an ice bath; it is used to determine the extent of winterization.

Cold test — related to winterization, cloud point, etc. It determines whether the oil will deposit crystals when stored in the home refrigerator.

Color — traditionally measured by viewing in a Lovibond comparator, but more recently by instrumental methods. Has an important effect on retail customer response.

Dilatometry — measures volumetric changes associated with temperature changes; detects phase transformations. One of the basic techniques used in determining the Solid Fat Index.

Free fatty acid test — measures the free fatty acid content of a sample of fat or oil calculated as oleic acid. Can provide an estimate of the degree of hydrolytic rancidity.

Iodine value — measures the degree of unsaturation in fats; involves determining the amount of iodine that reacts with a gram of fat.

Melting point — determines the liquefaction or solidfication temperature of a fat or oil; results are very dependent upon conditions of measurement and definitions of "liquid" and "solid."

MIU (moisture, insoluble unsaponifiable matter) — an indication of the amount of nonfatty impurities in crude oils and fatty acid products. Useful when pricing is based on the actual oil or fat in a batch of commercial material. Impurities may include moisture, volatile substances, insoluble matter, unsaponifiable matter, trace metals, and soaps of fatty acids.

Organoleptic testing — determinations of quality based on sensory clues, such as appearance, flavor, and texture are commonly used, especially in estimating purity (amount of contaminants), rancidity, degree of deodorization, and flavor reversion.

Oxygen absorption test — determines the stability of a fat by heating the sample in an oxygen atmosphere and measuring the rate of oxygen uptake.

Peroxide value — measures the extent of oxidation undergone by a fat; requires a determination of the amount of iodine liberated from potassium iodide by hydroperoxides (the oxidation products).

RI (refractive index) — a rapid purely physical test that can be used to estimate the iodine value of a sample to determine its type and source. Results are correlated with the extent of unsaturation in the fatty acids, and can be used to follow the progress of catalytic hydrogenation.

Saponification value — measures the amount of alkali required to completely neutralize the fatty acids hydrolyzed from the glycerides in a sample; it provides an indication of the average molecular weight of the fatty acids in a sample of oil.

Schaal method — estimates the oxidative stability of an oil by periodically organoleptically monitoring (smelling) samples held at 63°C. The end point is the length of time expiring before a rancid odor is detected.

Setting or congeal point — related to, but by no means identical with, the melting point; data are highly affected by conditions of the test, including particularly the rate of cooling and the definition of the end point.

Smoke point — determines the temperature at which a sample of fat emits thin, continuous wisps of smoke when heated under specified conditions.

SFI (solid fat index) — measures the solidity (or, usually, the specific volume) of a fat sample over a range of temperatures, and is related to the percentage of crystallline fat present in the sample. Traditionally based on dilatometric determinations, but more recently pulsed nuclear magnetic resonance has been used to give rapid measurements.

TBA (thiobarbituric acid) test — measures malonaldehyde, a by-product of oxidative rancidity; it is useful as a measure of oxidation only during the initial stage of oxidation.

Viscosity — related to ease of pumping, flow properties, cling in salad dressing, etc.

Sample Preparation

Although a substantial number of the samples presented to the analytical department of a large food manufacturer will consist of almost pure oils and fats (shortenings, frying oils, etc.), the majority of samples encountered by the analysts will probably consist of mixtures of fatty substances with several other food ingredients or components — bread, salad dressings, meat products, milk, etc. From these mixtures, the fat must be separated or extracted to give a more or less pure fraction that can be subjected to the techniques required to characterize and identify it and to determine its quantity relative to the weight of the original sample.

Many standard procedures for fat extraction have been developed, verified, and published. Most of these describe the preliminary treatment of the sample, such as methods for rendering it uniform throughout, for maintaining the integrity of the original components until the analysis begins (and afterward, for retention samples), equipment required for the analysis, conditions to be maintained during the analysis, steps involved during the actual determination, and numerical factors or constants to be applied to the raw data.

Extraction methods — Methods used to extract lipids from foods can be classified as:

(1) Reflux systems, exemplified by the decades-old Soxhlet apparatus, in which solvent vapors are condensed and flow through the sample container into a boiling flask from which come the vapors that are led to the condenser.

(2) Digestion by acids or alkalies prior to solvent extraction, such as the procedures that have been used for a long time in determining butterfat in milk samples.

(3) Non-heating methods — either dry column or solvent extraction.

Most extraction methods involve contacting the sample (often previously dehydrated) with a fat solvent (usually heated). The sample is often ground or milled after dehydration to give a finely divided material, provided the food or ingredient is not normally a liquid or paste. A solvent is chosen that will dissolve all expected types of lipids but in which most other substances (carbohydrates, proteins, etc.) are totally insoluble. A chloroform-methanol mixture has been used for dairy product, while diethyl ether has been widely employed for extracting cereals, meats, nuts, and confections. In many cases, hexane or a mixture of hydrocarbons has been used for cost and safety reasons.

The solvent is either circulated through a bed of the pulverized sample or is mixed with a relatively small amount of sample and thoroughly agitated until all the lipid has been dissolved. It is obvious that either of these alternatives will leave some solvent in the sample and some of the lipid will remain in the extracted sample after an equilibrium state has been reached if the suspended solid remains in contact with the solvent at the end of the extraction period.

This difficulty, which can be a minor source of inaccuracy in cases where an large excess of solvent is used and nearly all the solvent is removed by pressure or centrifugation, can be overcome by using percolation methods in which the hot solvent flows through a sample held in a porous container and is collected in a boiling flask from which vapors of the solvent are led to a condenser located above the solvent container. In such arrangements, the sample is continuously extracted by solvent that has been purified by distillation, while the soluble, non-volatile components collect in the boiling flask.

Effects of solvent — The fatty components collected in the solvent will depend on chemical and physical characteristics of the solvent used — its polarity, boiling point, miscibility with water and other solvents, solubility properties, and volatility. Non-polar organic solvents, such as hexane are suitable for most of the neutral or simple lipids (such as esters of fatty acids, acylglycerols, and unsaponifiable matter. Recently supercritical carbon dioxide has been used as a solvent for these simple lipids.

Complex or polar lipids, such as phospholipids, glycolipids, lipoproteins, oxidized acyl glycerols, and free fatty acids, are preferentially extracted by polar solvents such as methanol. These extractions require the breaking of ionic and hydrogen bonds formed between the lipid component and proteins or carbohydrates of the sample material. Solid phase extraction (SFE) is particularly useful for complex polar lipids (Mossoba and Firestone 1996).

Non-scientific considerations — Safety aspects must always be considered when developing or choosing a procedure. Unfortunately, many of the most effective solvents used are either flammable or toxic, sometimes both. Cost must also be taken into account, although the trend toward use of ever smaller samples has somewhat reduced the importance of solvent cost. Labor cost and disposal of used solvent or solvent fumes are factors seldom fully evaluated in designing these procedures, although their importance can hardly be exaggerated. Size of sample is usually not a major cost consideration, at least for the laboratory, though retention samples can ultimately take over large volumes of space that could better be utilized for other purposes.

Some Typical Analytical Procedures

It has long been recognized that the fat content of foods was related to the health of consumers and that purchasers of food were entitled to know how much fat was in the food they bought and consumed. In earlier times, the emphasis was on assuring the purchaser that the food had at least as much fat as was claimed or expected — a cookie or a cake labeled as being made with butter should have at least as much fat as was claimed and the fat had to originate from the desirable ingredient, butter. Fat was actually, in many cases, a desirable component, especially in times of food scarcity and for persons whose economic status gave them limited access to foods; probably fat, as fat, is still very desirable as a food in many parts of the world. As a condensed source of energy, it helps to maintain life in situations where an adequate supply of calories is difficult to obtain. Now, however, in the affluent, developed countries, the emphasis is on minimizing fat, and a claim of, say, 5% fat content is not at all satisfiable by including a little extra, "just to make sure."

Determining total fat content — For many years, total fat content has been measured for routine QC purposes by rather simple techniques involving extraction of a dried and ground food sample with a non-polar solvent such as hexane, or perhaps with diethyl ether or chloroform. The solvent is separated from the solid material, then evaporated and the dried residue is weighed. For samples in which some of the fat is highly emulsified and/or protected against solvents by non-lipid coatings, a preliminary treatment with acid has been used. There is no doubt that in complex mixtures where the fat may be bound to other components, or in which there are lipids that have acquired a polar nature because of their conjugation with other molecules, these straightforward techniques are not entirely satisfactory.

Rader et al. measured total fat and saturated fats in foods by utilizing packed column gas chromatography following acid hydrolysis. It was concluded that the acid hydrolysis-and-packaged column method satisfied the labeling requirements for determination of total fat and saturated fat for a large number of foods. Other authors reported a method for in situ preparation of fatty acid methyl esters for gas chromatographic analysis of fatty acid composition of food samples without a prior extraction step, but did not apply the method to determination of total fat.

Supercritical carbon dioxide extraction has many apparent advantages for the quality control laboratory — the solvent is not explosive or toxic, it is very inexpensive, and it does not ordinarily react chemically with sample components. Removal of the solvent from the extract is simple. It is not, however, as specific for fats as might be desired. Not only does it extract non-polar triglycerides, but also dissolves some polar components such as phospholipids, proteins, and carbohydrates. Notwithstanding the limitations of supercritical carbon dioxide extraction (which must be balanced against the defects of other methods), it has much appeal for the determination of the total fat content in food. Automated equipment is now available for solid phase extraction of food lipids with supercritical carbon dioxide.

Rapid and non-destructive testing with near-infrared reflectance spectrophotometry is feasible, and automated equipment is being offered for on-line testing for total fat content of foods. This method is reasonably accurate and, in most cases, fairly specific. Its greatest value is in the rapid, almost labor-free operation that it gives. Wide-line nuclear magnetic resonance has also been used for this purpose.

Separating, identifying, and measuring fatty acids — Once the mixed lipid sample has been obtained, further characterization involves the splitting of fatty acids from the other radicals, such as glycerol, that make up the compound lipidic molecules. A standard method for obtaining fatty acids from fats is saponification with alkali, a reaction which yields glycerol and alkali salts of the fatty acids. The latter are collected by acidifying the mixture, then partitioning the fatty acids into an organic solvent.

Transmethylation followed by gas-liquid chromatography of the fatty acid methyl esters is a common method for identifying fatty acids with carbon numbers 8 through 24, but it may not recover shorter chain fatty acids. Trans isomers of fatty acids can be measured by such instruments as infrared absorption spectrophotometers. When determinations of contents below 5% is needed, Fourier-transform infrared spectroscopy is often used. Individual *trans* fatty acids can be identified and measured by capillary column gas chromatography.

Mossaba and Firestone (1996) described a method attributed to House et al. that has undergone collaborative study and has been recommended by AOAC International as a first action method. This quantitative measurement of total fat in foods meets NLEA requirements. The principle is acid or alkaline hydrolysis of food samples followed by extraction with ether, then conversion of the fatty acids to their methyl esters (FAMEs) and quantitative measurement of the esters by capillary gas chromatography. This method is said to be applicable to foods containing 1% to 50% fat.

Determining the position of double bonds in fatty acids is a necessary step in establishing the structure of these compounds, and it has relevance to the nutritional role of fats. Methods include a combination of chemical and physical separation techniques. Among the recently developed techniques is a complex system adapting gas chromatographic and mass spectrometric equipment to the analysis of derivatives of fatty acids.

Ozonolysis methods require the reacting of ozone with fatty acids or fatty acid derivatives so as to split the fatty acid molecule at the unsaturation points, resulting in stable compounds that can be separated by gas chromatography. Hydrazine reduction is yet another procedure that has been successfully used.

Chromatographic techniques — Separation and identification of lipids and fatty acid derivatives can be performed by chromatographic techniques such as thin layer chromatography, size exclusion chromatography, and high-performance liquid chromatography.

In thin layer chromatography, aluminum glass plates are coated with an adsorbent that has been selected for the compatibility of its properties with the desired application; for example, silica gel G containing calcium sulfate is often used for separating cholesterol and cholesterol esters, mono-, di- and triacylglycerols, free fatty acids, and phospholipids. Silica gel with silver nitrate is used to separate fatty acids according to their degree of unsaturation and according to the cis- or trans-configuration of the double bonds. Plates are developed in a closed container with an appropriate solvent system. The localized concentrations of solutes that occur as a result of separation during the migration of the solution can be made visible by spraying the plate with a fluorescent dye and then illuminating the plate with ultraviolet light. Other methods are also being used. Micro-column thin layer chromatography is based on the use of quartz rods coated with silica gel-glass frit.

Size exclusion chromatography employs a stationary phase consisting of cross-linked macromolecules (such as copolymers of styrenedivinyl benzene) that allows smaller molecules to diffuse mechanically into pores of appropriate size while excluding larger molecules. It has been used successfully to separate the fat-degradation products that accumulate in frying fats.

High performance liquid chromatography can be used to separate nonvolatile, high molecular weight lipids in several instrumental modes based on either partition or adsorption chromatography. Mono-, di-, and triacylglycerols, sterols, and free fatty acids can be separated using a gradient of ethanol into a 9:1 mixture of hexane and chloroform with an adsorbent element consisting of silica gel. Location of solutes is typically determined by detectors operating in the ultraviolet-visible range, but refractive indexes, flame ionization, and light-scattering (mass detectors) have also been used.

Gas chromatography is an effective method for separating and quantifying lipids and fatty acid derivatives such as the methyl esters. Good resolution of mixtures of fatty acids into their components of different chain lengths and degree of saturation is usually achieved achieved using capillary columns with highly polar liquid phases, but success of these systems is not as good when isolation of geometric isomers or positional isomers is desired.

Capillary gas chromatography combined with Fourier-transform infrared spectroscopy is said to allow on-line generation of spectra, allowing double-bond configurations to be confirmed for individual geometric isomers in complex mixtures of saturated and unsaturated fatty acid methylesters. The analyst can simultaneously monitor both the chromatographic peaks and the Fourier-transform spectra of mixture components as they are eluted.

Supercritical fluid chromatography uses supercritical carbon dioxide as the mobile phase. It includes features of both capillary gas chromatography and high performance gas chromatography. Operations can be carried out near room temperature, thus reducing heat degradation of compounds of interest. Flame ionization detectors are often used to sense the eluting substances.

Other analytical methods — Nuclear magnetic resonance equipment has been used in the fats and food oils industry to monitor processes and for quality control purposes. It can rapidly determine the oil contents of seeds and meals, the solid fat content of fats, and total fat content of foods. Methods of this type have been sanctioned by the American Oil Chemists' Society and are available for determining the oil content of rapeseed and the solid fat content of commercial shortenings.

EXAMPLES OF PRODUCTS WITH DECREASED FAT CONTENTS

There have been hundreds, if not thousands, of formulas or recipes developed by vendors of specialized dietary food ingredients that have appeared in the literature (including patents). Only a small selection of these can be provided in the space available. The reader is cautioned that some of the foods resulting from the formulas may not have been adequately or objectively tested, especially for organoleptic characteristics such as flavor, appearance, texture.

Whipped Cream Replacers

Whipped cream is a widely liked topping for pies, especially cream pies, ice cream sundaes, cakes, and other types of desserts, and it is often used on fruit preparations such as strawberry shortcake and the like. Sometimes it is used as a topping on hot chocolate or coffee. On a weight basis, whipped cream is a very high calorie food, but on a volume or per-serving basis it is a much more reasonable dietary addition. Probably the display or appearance features provided by whipped cream are as important to most consumers as the flavor, which is bland, or eating texture, which is smooth but otherwise lacks character.

There are many whipped cream substitutes in the typical super-market; some of the best types, offered frozen in plastic tubs or chilled in pressurized dispenser cans, are difficult to distinguish from real whipped cream. Powdered bases for addition to water and then whipping into a sem-blance of whipped cream are also available for manufacturers of retail products, food service operators, and consumers.

A fluid mix for whipped topping has been made according to the following formula:

> 0.50% Avicel RC-581 stabilizer
> 60.90% water
> 10.00% sugar
> 2.00% sodium caseinate
> 18.00% hard fat
> 8.00% mono- & diglycerides
> 0.60% lactic acid propylene glycol ester
> Flavors and colors may added according to preference.

Processing is as follows: Add stabilizer to recommended amount of water. Mix thoroughly and heat to 90°F. Dry mix sugar and sodium casein-ate, then add them to the remaining water. Heat to 110°F and add the remaining ingredients. Heat this mixture to 160°F for 30 minutes, then homogenize it at 1,000 and 500 psig through a two-stage homogenizer. Cool the product as rapidly as possible to 40°F, and hold it at 40°F for at least twelve hours before whipping.

Peanuts and Peanut Butter

The history of experiments into defatting peanuts so as to provide a relatively low calorie snack extends back probably 40 or 50 years. The usual approach, which has been moderately successful from a technical standpoint but not very successful from a consumer acceptance standpoint, consists of first pressing whole or half peanuts (usually blanched) to remove a large part of the oil, hydrating the distorted peanuts so as to allow them to recover most of their unpressed volume, then roasting to develop the typical peanut flavor.

For a recent example, we refer to the patent of Zook (1992), which teaches the pressing of peanuts to remove 40% to 52% of the oil from blanched or unblanched shelled nuts, contacting the shrunken kernels with about 4 pounds of water per 100 lbs of nuts (steam can be used), roasting the kernels under well known conditions, and coating the roasted nuts with less than about 10% of an edible oil such as corn, cottonseed, peanut,or coconut oil. The inventor recommends mixing the defatted product with full-fat roasted peanuts before packaging. She claims a net decrease in calories of between 8% and 15%, and a net decrease in fat of between 15% and 25% for such a mixed product.

Peanut butter is a difficult material to simulate with low-fat imitations. The smoothness and spreadability that constitute a large part of its appeal is hard to reproduce in a product having substantially lower lipid content. Furthermore, the use of water-based diluents creates severe problems due to absorption of the water by the non-fat solids of peanuts, so that adverse textural properties and undesirable flavor notes become evident. Also, the water activity change resulting from use of aqueous additives creates a susceptibility to microorganism attack that is not present in regular peanut butter.

These obvious problems have not deterred entrepreneurs from making an effort to satisfy consumer demand for a low-fat peanut spread. For example, the patent of Lasdon et al. (1989) describes a low-fat low-calorie peanut spread made by milling and cooking defatted peanut flour with water. Their formula and procedure are basically as follows:

Formula
  9.00 lb peanut flour
 13.50 lb water
  6.75 lb corn syrup
  1.875 oz molasses
  0.57 oz xanthan gum
  0.86 oz lecithin
 12.00 oz sugar
  0.93 oz coloring
  0.54 oz potassium sorbate
  2.40 oz flavorings

Process
The peanut flour (commercial examples of which may be obtained with fat contents from 1% to 35%) is mixed with water and the rest of the ingredients and milled to particles having a maximum size of 75μm. This mixture is then cooked at a temperature above 175°F but below boiling for about 45 minutes. After cooling to no lower than 170°F, the product is filled into jars. It requires refrigeration for extended shelf life.

Results
The finished material has the fat content reduced by as much as 80% to 95% compared to regular peanut butter, i.e., to 1% to 35% fat content, depending on the additives and the water percentage, It has a texture and a flavor similar to those of peanut butter.

Franklin (1994) described a somewhat different method of making reduced fat peanut butter. In his invention, oil is removed from ground peanuts until a supernatant oil layer is formed. Then, the peanuts are removed from this oil layer and may be reground. Maltodextrin and modified food starch are added to the peanut paste, either before or after the optional regrinding. Salt, butter, flavoring, and/or peanut flavoring may also be added to and blended with the separated peanuts.

Tortillas

Although the original masa tortilla contained only a small amount of fat, none of it added as an ingredient, present versions of both masa and wheat flour tortillas have a fat content of a few percent. Home produced tortillas normally contain less fat that do the commercial varieties, the latter requiring shortening to maintain flexibility and a relatively soft texture during distribution and storage. Even so, they do not have particularly long shelf lives, although they may be usable for longer periods of time than white bread, when stored under comparable conditions.

As a result of the ingredient fat additions (between 5% and 15% of the flour weight), commercial versions of tortillas tend to be rather high in calories, and there is an understandable desire to be able to claim low-fat status for the items. Some improvement can be achieved by using fairly low additions of monoglycerides or other good dough conditioners. Gums, such as modified starches, carboxymethylcellulose, xanthan gum, gum arabic, and various kinds of fibers have been used, often in combination with increased water.

High moisture-low fat tortillas have been made with pea hull fiber (Best 1996). The developer, Gary Nickel, said, "We found that tortilla doughs formulated with customized blends of our pea fibers retained their machinability even at much higher moisture levels. Whereas traditional tortillas may retain 29 to 31% moisture, our non-fat tortillas can go as high as 43% moisture and still maintain their flexibility and shelf life properties." There is a calculated content of 4.5 to 5.0 grams of dietary fiber per serving (two tortillas).

Masa and doughs with added fat are easier to process. A major problem in the production of wheat flour tortillas made by the hot-press method, is that a resting period is needed so that the dough can relax before it is pressed into the circular sheets typical of the product. Short rest periods lead to dough pieces that are difficult to process and give tortillas of inferior quality. Reducing agents such as L-cysteine, bisulfites, sorbic acid, and fumaric acid are often added to shorten the rest time and improve other processing characteristics of the dough.

A low fat tortilla formula was developed by technicians at Lallemand,

Inc. The ingredient "Fermaid T" mentioned in the following formula is a dough conditioner distributed by that company.

Formula
100 parts flour
  58.66 parts water
    0.67 parts shortening
    0.6 parts sodium stearoyl lactylate
    2.0 parts Fermaid T
    1.67 parts salt
    0.83 sugar
    0.19 gum
    2.67 baking powder
    0.67 sorbic acid

Process
  Mix all ingredients to form a cohesive dough. Temperature should be 30°C. Divide dough into round pieces of 55 grams. Rest dough pieces for 12 minutes at 30°C. Press dough pieces to form discs. Temperature of the pressing plates should be 400°F. Bake tortillas at 362°F. Cool and package.

  A comparable tortilla formula made with a typical amount of shortening, would contain 10 parts of added fat (FWB), according to this publication.

LABELING AND OTHER REGULATORY CONSIDERATIONS

  The technological challenges posed by fat replacement formulas are but a small part of the obstacles confronting the developer of reduced fat products. When synthesized fats are being used, new products face present or potential problems with legal clearance and labeling, and some may never be approved. If the fat reduction is being accomplished by substituting some food ingredient that has been used for a long time, perhaps in conjunction with added water, the problem is somewhat simpler. Even then, federal regulations must be strictly followed.

  Fat has been defined by the FDA, for nutrition labeling purposes, as the sum of fatty acids expressed as triglyceride equivalents. Probably 95% of dietary lipids (fats and oils) are triglycerides, compounds made up of three fatty acids plus glycerol. The FDA definition of fat takes into account the fatty acids in triglycerides was well as in the other dietary lipids, the latter usually being regarded as including monoglycerides, diglycerides, free fatty acids, phospholipids, glycolipids, and sterols such as cholesterol. The Nutritional Labeling and Education Act specifies that the number of grams per serving of total fat and the percent daily value (%DV) of total fat must be listed on the labels of packaged foods. In addition, the grams and %DV of saturated fats (the sum of all fatty acids containing no double bonds) must be declared. Listing of polyunsaturated fats (FDA definition = *cis, cis*-methylene interrupted polyunsaturated fatty acids) and mononunsaturated fats (FDA definition = *cis*-monounsaturated fatty acids) is at the option of the packager.

The NLEA did not deal in detail with *trans* isomers. These compounds cannot be included in voluntary listings of unsaturated fats and will not be listed separately on new food labels. According to the labeling regulations that became effective in May 1994, the FDA stipulated that *trans* fatty acids could not be included in the voluntary listing of unsaturated fatty acids, and it imposed a limit on *trans* fatty acid content of products for which saturated-fat-free claims were made

The FDA has restricted the use of certain nutrient descriptors pertaining to foods ostensibly reduced in fat content as compared to previous versions of the foods. Among the approved terms are:

Free — less than 0.5 gram of total fat per reference amount of food and per labeled serving (or for meals and main dishes, less than 0.5 g per labeled serving). No ingredient that is fat or understood to contain fat, except if the ingredient listed in the ingredient statement has an asterisk that refers to a footnote, "adds a trivial amount of fat."

Healthy — the food must be low in fat and saturated fat, and a serving must contain no more than 480 mg of sodium and no more than 60 mg of cholesterol. cholesterol.

Less or *Reduced* — at least 25% less fat per reference amount than an appropriate reference food; reference food may not be "Low fat."

Light — the altered product contains one-third fewer calories or 50% of the fat in a reference food; if 50% or more of the calories come from fat, the reduction must be 50% of the fat.

Low — contains 3 g or less per reference amount, and per 50 g of food if the reference amount is small.

Percent fat free — the product must be low-fat or fat-free, and the percentage must accurately reflect the amount of fat in 100 g of food. Thus 2.5 g of fat in 50 g of food justifies in a "95% fat-free" claim.

POSSIBLE RISKS IN EXCESSIVE REDUCTION OF FAT

Not all nutritionists are convinced that lowering the fat content of the diets of normal individuals to a very low level will in all cases improve their health, longevity, and quality of life. Siguel and Lerman (1996), in clinical studies, found that low-fat diets using processed low-fat and nonfat foods deprived of essential fatty acids decreased high-density lipoprotein levels and increased triglyceride levels. Some of their patients felt more tired and complained that they lacked energy. These reearchers warned that such low-fat diets, when used as in maintenance regimens (i.e., without weight loss) increase the ratio of total cholesterol to HDL-cholesterol, and deplete essential fatty acids in the plasma and red blood cells. Furthermore, they cite studies showing that low levels of polyunsaturated fatty acids are associated with increased risk of heart disease. Siguel and Lerman claim that the lipid values of these patients are particularly sensitive to deficiencies in polyunsaturated fatty acids, and conclude that low-fat diets that are also low in polyunsaturates are dangerous for many people and may be deadly for patients with low HDL and high triglyceride levels.

Other writers have revealed that reduced levels of omega-3 polyunsaturates in red blood cells are associated with increased risks of cardiac arrest. Low HDL and high triglyceride levels are probably the most common genetic abnormality of lipid metabolism, reputedly a common cause of premature death, and a major factor in heart disease costs.

Furthermore, demographic studies lead to questions about the health significance of saturated fats and the importance of the ratio of fat calories to total calories in the diet. One commentator pointed out, "Compared with whites, Korean-American men have a five-fold higher incidence of stomach cancer and an eightfold higher incidence of liver cancer . . . Asian Indians have one of the highest rates of coronary artery disease, despite their largely vegetarian diet, which would limit cholesterol and saturated fat intake. [but] Risk factors such as high serum cholesterol levels, cigarette smoking, and hypertension are not disproportionately present in Asian Indians. However, lipoprotein α-1, an independent risk factor in the development of premature atherosclerosis, is elevated disproportionately in Asian Indians. If lipoprotein α-1 levels are indeed a strong genetic predeterminant of CAD in Asian Indians, health programs might be tailored to specifically lower these levels."

chapter five

ENGINEERED FATS AND FAT MIMETICS

As explained in another chapter, reduction of fat in products that conventionally contain substantial amounts of that type of ingredient can be accomplished by (1) simply reducing the amount of fat, keeping the other ingredients in their usual ratio, and accepting the decline in quality that is likely to accompany such changes, (2) using hydrocolloids or the like to bind water in gel-like structures that simulate the unguentary and tenderizing effects of shortenings, (3) using triglycerides that have been selected or fractionated or synthesized so as to provide lower calories while still retaining most of the properties of common fats and food oils, and (4) synthesizing products that have some of the properties of fats but which are not absorbed by the body, or if they are absorbed, yield much less energy per unit weight.

It is the last two categories with which we will be concerned in this chapter. These two groups of ingredients are clearly differentiated. Engineered fats contain only normal fatty acids, i.e., unbranched carbon chains with a carboxyl group at one end and a methyl group at the other. They are found in nature, though not always widely distributed or in large amounts, and the ingredients offered to bakers may have, in fact, been synthesized in commercial equipment. In the other group, fat mimetics may be based on a fatty acid chain, or they may not, but, if so, the molecule has been modified by adding branches, atoms not found in normal fatty acids, and chemical groups foreign to normal fats. Most are not found anywhere in nature, while other examples bear very little chemical resemblance to triglycerides of fatty acids.

ENGINEERED FATS

Engineered fats can be defined as triglycerides that have been fractionated or synthesized so as to have different properties than the fats and edible oils that are familiar items of commerce. They are natural in the sense that all of the molecular species can be found somewhere in nature, though not necessarily in any great amounts or not in the purity desired by the dietetic food industry.

In the category of engineered fats we find, for example, the triglyceride esters of short-chain fatty acids. Long chain fatty acids, such as are found predominantly in nature, have more energy content per gram than short chain fatty acids because their carbon to oxygen ratio is higher. Thus it is logical that fats containing mostly short-chain fatty acids would have a lower calorie content than fats containing mostly long-chain fatty acids, at least according to bomb calorimeter measurements.

Very Short Chain Fatty Acids

The recognition that lengths of the three fatty acid residues making up part of the triglyceride affect the caloric content of the compound has led to the development of synthetic fats containing very high proportions of very short chain fatty acids. A triglyceride in which all three of the fatty acids are of very short chain length can be expected to have physical properties, such as melting point, significantly different from the fats and oils usually found in nature. A closer approach to the characteristics of commercially familiar lipids can apparently be achieved by causing at least one of the fatty acids to be of the long-chain type.

The generic name "salatrim" (an acronym of the term Short- and Long-chain Acid Triglyceride Molecules) has been assigned to this group of lipids, and a tradename "Benefat" has been applied to some of the commercial versions. Benefat #1 contains the short-chain fatty acids acetic, propionic, and butyric acids and the long-chain stearic acid, and perhaps others in minor amounts; and it is said to have an energy content of five calories per gram.

Benefat has been used in chocolate chips for baking, for which it has the additional advantage (it is said) that tempering of the chocolate before molding is not required. This eliminates critical processing steps required in the production of real chocolate products.

Similar in concept is a manufactured lipid called caprenin. This follows a somewhat similar principle to the one mentioned above, however the specific fatty acids in caprenin are caprylic (C8), capric (C10), and behenic (C12), medium-chain acids. It was originally created to mimic cocoa butter rather than as a low-calorie fat.

The calorie content, so far as the human digestive system is concerned, will depend on the specific fatty acids used. Short chain fatty acids will yield less energy than longer chain varieties because of their lower carbon to oxygen ratio, while the long-chain fatty acids that are used in the commercial salatrims are said to have been selected because they are not fully absorbed. Some forms of salatrim are said to contribute only 5 calories per gram to the diet, as opposed to the average of about 9 calories per gram contributed by the usual natural fat ingredients.

The triglycerides making up salatrim are broken up in the digestive tract to give fatty acids and glycerol, as occurs with other fats, so they could presumably be regarded as safe as any other fat. The manufacturer petitioned the FDA to grant GRAS status for Salatrim in December of 1993, and it was accepted for filing in June of 1994.

It is to be expected that the physical properties and some of the chemical properties, if not the nutritional characteristics, of fats made up largely of very short chain fatty acids would differ substantially from those characteristics of fats normally encountered in human diets.

Low-fat Replacements for Cocoabutter

Designing low-fat replacements for candies and the like that are customarily based on the fatty structures provided by cocoabutter is unusually difficult. The smooth, melt-in-the-mouth texture of solid pieces of sweet chocolate, semisweet chocolate, and milk chocolate is almost impossible to duplicate with any nonfatty substance and is, in fact, very difficult to simulate with other kinds of fat. Baking chips (such as are used in chocolate chip cookies) of a reduced fat type also pose serious formulation challenges but the cookie matrix in which they are consumed tends to obscure some of the texture problems and often some of the flavor differences as well.

The formula which follows (Kosmark 1996) is said to provide a chip containing 23.95% fat — 3.10% of fat at 9 calories per gram and 20.85% of the modified fat "Salatrim" — having a caloric density of 5 calories per gram. It is believed that a similar formula was used in making the "Reduced Fat Baking Chips, Semi-sweet Chocolate Flavor," recently introduced by Hershey. This product is said to contain 50% less [calories from] fat than the company's regular Semi-sweet Chocolate Chips.

Formula for Reduced Fat Baking Chips
62.20% sugar
20.85% Salatrim (Benefat #1)
15.50% cocoa, natural, 10%-12% fat
1.00% milk fat, anhydrous
0.40% lecithin, fractionated
0.05% vanillin

The author previously quoted (Kosmark 1996), reported that taste tests were conducted comparing commercial chocolate-coated raisins (pan-coated) with a lower calorie product the coating of which contained a large amount of Salatrim. The experimental product was said to have equalled the control in preference ratings. Also, coatings for frozen dairy bars have been developed that lead to as much as a 33% reduction in calories from fat. Research work on caramel, toffee, and cremes containing salatrim is said to be in progress.

Formulas and processes for milk chocolate flavored and dark chocolate flavored confectionery coatings using Benefat as a replacement for some of the cocoabutter have been published by Venardos (1996) and are reproduced in Table 5.1.

Table 5.1
CHOCOLATE-FLAVORED CONFECTIONERY COATINGS[1]

Ingredient	Milk chocolate	Dark chocolate
Part 1: Kneader/mixer		
Sugar	53.54	57.04
Benefat #1	13.50	12.60
Nonfat dry milk	9.00	--
Cocoa 10/12 fat natural	5.00	14.50
Cocoa 10/12 fats alkalized	2.00	--
Anhydrous milk fat	1.00	0.90
15% PC lecithin	0.10	--
40% PC lecithin	--	0.10
Vanillin	0.03	0.05
Veltol	0.03	--
Part 2: Finished mix.		
Benefat #1	15.50	14.51
15% PC lecithin	0.30	--
40% PC lecithin	--	0.30
Nutritional information[2]		
Total fat, grams[3]	12	12
Available fat, grams	7	7
Total Calories	160	160
Calories from fat	70	60

[1] Amounts are stated on a parts basis.
[2] Nutritional information is for a forty-gram serving.
[3] A typical full-fat coating contains 14 grams total fat per 40 gram serving.
Source: Venardos (1996).

Processing

The Part 1 ingredients were combined in a kneader/mixer by methods customary in the chocolate industry. The material was refined to a particle size of between 30 and 33 microns. The coating does not need to be tempered and can be held at a temperature as low as 95°F or as high as 120°F prior to enrobing, panning, or depositing. The coating sets up much quicker in the cooling tunnel than is the case with standard milk chocolate coatings. Due to the rapid rate of crystallization, all vessels and transfer piping need to be heat-jacketed.

Medium-chain Triglycerides

Manufactured fats containing predominantly saturated fatty acids with chain lengths of 6 to 10 carbon atoms have been referred to as MCTS, medium-chain triglycerides. The stepwise process for synthesizing these materials consists of the hydrolysis of vegetable oils, fractionation of the fatty acids to obtain a concentrate of the 8- and 10-carbon compounds caprylic and caproic acids (usually, with some 6-carbon), and re-esterification of the concentrate with glycerol to form triglycerides.

There is not a universal agreement amongst oil chemists as to the meaning of the term, "medium-chain triglycerides." Some have described the oil as having 10 or fewer carbon fatty acids, others say the range is between 6 and 12 carbon fatty acids (see for example, the correspondence of Jones 1991 and Babayan 1991).

In whatever way they are defined, these preparations have low viscosity and bland flavor, are colorless, and are extremely resistant to oxidative rancidity. Most of the present discussion is based on the article by Megremis (1991).

According to published data, the resistance to oxidation is remarkable as shown by the following figures in which the "hours" refers to the duration of the induction period:

Fish oil, 0.3 hr
Sunflower oil, 6.7 hr
Soybean oil, 11.2 hr
Canola oil, 14.7 hr
Mineral oil, 18.0 hr
High-oleic sunflower oil, 25.3 hr
Olive oil, 27.3 hr
Hydrogenated soybean oil, 160.0 hr
Medium-chain triglycerides, 180.0 hr

MCTS are metabolized differently than the more common longer chain triglycerides, partly because these fats are more water dispersible than the longer chain species. When ingested, ordinary fats are hydrolyzed into long chain fatty acids and monoglycerides, then absorbed, re-esterified, incorporated into chylomicron structures, and transported into the lymphatic system of the body. The MCTS, however, are rapidly and completely hydrolyzed, yielding relatively small molecular free fatty acids that are absorbed through the membrane of the small intestine, then are bound to serum albumin and travel from the intestine to the liver by way of the portal vein.

MCTS contain about 10% fewer calories than most triglycerides found in animal and vegetable fats, as determined by bomb calorimetry. In addition, the body is inefficient in converting MCTS to body fat, so that the net calories they provide are actually lower than the 10% reduction predicted by bomb calorimetry.

MCTS have some practical problems, however. When used in frying, they have a distressing tendency to burst into flame. They can also impart undesirable flavors reminiscent of coconut, a rather typical symptom of rancidity. These fats are not recommended for diabetics because of the ketone-body buildup and decreased cell mass of the liver that may occur in persons suffering from that disorder.

Yang (1989) patented a type of MCT that is said to overcome some of the negative features of these lipids while retaining the desirable low caloric response of the class. Briefly, the reduced calorie fat includes synthesized combinations of saturated medium chain, saturated long chain, and unsaturated long chain fatty acids. The inventor gives a formula for mayonnaise incorporating the improved MCT: 75% of the patented substance, 10% vinegar, 9% egg yolk, 3% sugar, 1% salt, and 2% mustard.

FAT MIMETICS

One of the earliest patents for a fat mimetic took a very straightforward approach, disclosing the use of white (purified) mineral oil as a food ingredient to replace fats and food oils. This material is a mixture of liquid hydrocarbons obtained from petroleum. Although hydrocarbons will show a very high energy content by bomb calorimetry measurement, or the like, they cannot be utilized by the human digestive system, rating zero calories. Sounds good, but the unfortunate side effects of these compounds prevent their use in foods. They elicit typical foreign body reactions in the intestinal mucosa, lymph nodes, liver, and spleen, and they interfere with the absorption of fat-soluble vitamins, such as vitamin A. Mineral oil has been used, is probably still being used to some extent, as a lubricant in food processing to prevent sticking, and as a laxative.

The patent literature contains many examples of synthesized compounds that have been developed for possible use in foods, according to the expressed intent of the inventors. This group of fat replacers, sometimes called fat mimics, fat mimetics, or fake fats include sucrose polyesters, polyglycerol esters, triglyceride esters of alpha-substituted carboxylic acids, alkyl glycoside polyesters, and several other types of chemical compounds. None of these are triglycerides of fatty acids, and so they can be expected to differ in numerous respects from true fats. Many of them have not yet been approved for food use, and some of them may never achieve this status.

One of the few substances in this category to have advanced to commercial acceptance are the sucrose polyesters. Among these is "Olestra," the trade name for a mixture of sucrose molecules that have been chemically bonded to six, seven, or eight long chain fatty acids. Olestra closely resembles ordinary cooking fats and oils in appearance, texture, and lack of flavor, according to Raber (1995). It is not digested in, or absorbed from, the human intestinal tract, so it contributes no calories to the diet.

Olestra, and presumably other sucrose polyesters, are differentiated from many of the other fat substitutes by their fairly good stability at high temperatures, such as those encountered in baking or frying. Other modes of use have been listed by Nelson (1990).

There appear to be some negatives to Olestra's use in foods; critics say the substance causes abdominal cramping and diarrhea. It also inhibits the absorption of some vitamins and other nutrients in the digestive tract. The effect of the relatively very stable Olestra on the environment as it accumulates in waste, and in particular its effects on water quality and wastewater treatment processes, has apparently not been examined in any detail (Giese 1996).

Acylated glycerides of a specific chemical structure were suggested as ingredients in low-calorie fat-containing food products in the patent of Volpenheim (1986). The inventor states that these acylated glycerides have a substantially lower effective caloric value because they are less efficiently digested and absorbed in the intestinal tract than ordinary (triglyceride) fats. Proof of resistance to pancreatic enzymes in vitro was given in the patent, but no tests showing the caloric content when ingested as a food can be found in that publication. An exemplary formula for cake mix (to be added to water) containing acylated glyceride of the kind covered by the patent was given, as follows.

36% cake flour
44% sugar
13% glycerol tri-alpha-oleoyloxy palmitate
 4% nonfat dry milk
 2% leavening
 1% salt

Foods Made with Fat Mimetics

The patent literature is replete with examples of formulas for products made with fat mimetics. The reader is cautioned that these formulas do not necessarily make marketable, or even edible, products, and they should be recognized as merely starting points for further development. In addition,

FDA clearance for use of these materials in foods may not have been obtained and, in fact, approval may not even have been sought by the patent holder. Nonetheless, the formulas are often informative and may be very helpful if FDA approval for a particular mimetic is eventually obtained.

Polyvinyl alcohol fatty acid esters — These compounds, and particularly the esters made with unsaturated fatty acids, constitute a relatively new class of edible fat replacements. Polyvinyl oleate appears to be the preferred compound. For further information, see the patent of D'Amelia and Jacklin (1990). As an example of its use, the following formula for a cookie is given.

(1) Cream together 23.11 parts fine granulated sucrose, 0.37 parts salt, 0.44 parts sodium bicarbonate, and 11.48 parts polyvinyl oleate. "At this stage, no 'foam' appears as with solid shortening during creaming." Add to this mixture 3.74 parts distilled water and 5.87 parts dextrose solution (5.93%), then finally 39.10 parts flour. The very tacky dough that results is either extruded or rolled out and cut to make cookie pieces that are baked typically at 400°F for ten minutes. When prepared according to these directions, the finished cookie should contain less than 1% fat, which is less than one-tenth the amount expected in a cookie made with a normal percentage of shortening.

Sprayed crackers were also described by these inventors. Prepare a dough from 100 parts flour, 5 parts sugar, 1.5 parts malt, 7.5 parts of polyvinyl oleate, 1 part salt, 0.9 parts sodium bicarbonate, 2.5 parts nonfat dry milk, 2.5 parts high fructose corn syrup, 0.75 parts monocalcium phosphate, and 28 parts water. The dough is sheeted, stamped, and sprayed with polyvinyl oleate, then baked. This seems to be an extraordinarily dry dough. The reduction in fat is not entirely clear, but seems to be about 80% less than a normal fat-containing, fat-sprayed cracker.

Polyvinyl oleate does not appear to have been cleared for food use at the time of this writing.

Thioester derivatives — These fat mimetics have a two- to six-carbon backbone to which is attached by an ester linkage at least one C_1 to C_{29} fatty aliphatic, ether, ester, or thioester group, plus at least one other fatty C_1 to C_{29} group in ester or thioester linkage. The best of these materials have three fatty substitutes attached to the backbone and are partly utilizable by humans. In the patent of Klemann et al. (1991), are disclosed several examples of bakery products incorporating types of thioester derivatives as fat replacements.

Low calorie [imitation] sweet chocolate can be prepared by mixing one part of cocoa powder (fat content not specified), one part sugar, and one part of a specific thioester described in the patent. After the fat, the cocoa powder, and part of the thioester have been thoroughly blended, they are passed through a refiner to reduce the particle size. Then the material is conched, and the remaining thioester is added. The mixture is poured into molds and quench cooled. Unlike real chocolate, no tempering sequence appears to be used before molding.

Sugar cookies can be prepared by blending 231 parts of sugar, 114 parts of a specific thioester described in the patent, 3.7 parts salt, 4.4 parts sodium bicarbonate, 37.4 parts of water, 58.7 parts of a 5.9% dextrose solution, and 391 parts of flour. The dough is formed by extrusion and baked by the usual process.

Current legal status of these materials is unknown, but they are apparently not being used in commercial food products.

Sucrose polyesters — Compounds of this type have been discussed previously in this chapter, but further details on other variations can be found in the patent of Mijac and Guffey (1989), who disclosed a synthesizing method for sucrose polyesters of a specific kind, and gave examples of their use in food products. Two examples are given below.

Shortcake

Formula: 1.75 cups of sifted all-purpose flour, 1 teaspoon salt, 3 teaspoons double-acting baking powder, 1 tablespoon sugar, 4 to 6 tablespoons sucrose fatty acid ester shortening, and 1 cup milk.

Method: Cut shortening into sifted dry ingredients. Add milk to a depression in the center. Stir until the dough cleans the side of the bowl. Turn the dough onto a floured board and knead, making about 8 to 10 folds. Roll with a lightly floured rolling pin to get the desired thickness. Cut into desired shapes with a lightly floured die, and place them on an ungreased baking sheet. Bake until lightly browned.

Puff pastry

Formula: 1 cup sucrose fatty acid ester shortening, 1.5 cups all-purpose flour, 0.5 cup sifted cake flour, 0.25 teaspoon salt, and 0.75 cup cold water.

Method: Two tablespoons of the shortening are reserved, and the remainder is formed into a sheet by pressing it between waxed paper, then refrigerating it for later use. Combine flour and salt in a bowl, work the 2 reserved tablespoons of shortening into the flour by rubbing. Add water and stir until the dry ingedients are moistened. Then knead 30 times on a floured board. Let the dough rest 10 minutes, then roll it, forming a circle. Place sheet of shortening on the dough and cover it by folding over edges of dough. The usual sequence of resting, rolling out, and refolding that is typical of puff pastry processing is repeated as many times as necessary to get the layering and flakiness desired in the finished pastry. Form or cut into desired shapes and bake at 500°F for 5 minutes, reduce heat to 375°F and bake for about 30 minutes more.

Another inventor describes the use of sucrose polyesters synthesized to behave like cocoabutter (McCoy et al. 1989). The compounds patented are sucrose fatty acid esters having at least four fatty acid ester groups, and having the following fatty acid composition: (a) from about 25% to about 50% lauric acid, (b) from about 50% to about 75% stearic acid, and (c) up to about 5% other fatty acids. These materials are noncaloric and are said to be less expensive than natural cocoa butter. A chocolate product was made according to the following recipe: 48% sugar (12X), 14% chocolate liquor, 18% milk solids, and 20% of sucrose fatty acid ester cocoa butter substitute. It appears from the processing instructions that imitation chocolate prepared in this manner must be tempered in a sequence similar to that used for real chocolate. The inventors claim that the imitation chocolate is good-tasting and melts rapidly in the mouth.

Considerable thought has been given to the possibility of using the imitation, non-caloric (or low calorie) fats as frying oils, but many of the fake fats cannot be used in frying since they do not have the physical properties required and will break down rapidly under the high temperature conditions encountered in frying vats. It appears, however, that Olestra, or some forms of Olestra, will function satisfactorily as heat exchange mediums. Obtaining FDA clearance for the material for this use has been a long-term project but, as this is being written, announcements have been made (McGinley and Narisetti 1996; Allen 1996C) telling us that Procter and Gamble's Olestra had received approval for use in fried savory snacks; it is not yet accepted for use in any other fried foods, including sweet snacks.

The substance (Olestra) may be used "in place of fats and oils in prepackaged ready-to-eat savory [i.e., salty or piquant but not sweet] snacks. In such foods, the additive may be used in place of fats and oils for frying or baking, in dough conditioners, in sprays, in filling ingredients, or in flavors" (21 CFR 172, in a new section 172.867, published January 30, 1996).

Reports indicate this imitation fat can replace up to 100% of the conventional oils and fats used in the preparation of savory snacks such as chips of various kinds, crisps, extruded snacks, and crackers. Considerable reductions in calorie content can be achieved by such replacements, for example, it has been stated that potato chips fried in Olestra have a calorie content of about 50 per ounce as compared to 70 calories for an ounce of regular fat-fried potato chips. An ounce of regular tortilla chips will have, on average, about 140 calories while tortilla chips fried in Olestra will show about 90 calories per ounce (Giese 1996). These are certainly worthwhile reductions in energy content for the snack-loving dieter.

Although Olestra has not yet been approved for use in fried sweet bakery products, it is probably only a matter of time before such approval will be sought and granted, if problems do not arise from its currently approved uses. A major question is whether doughnuts fried in Olestra will taste the same, appear the same, and have the same texture as doughnuts fried in animal or vegetable shortening; there do not seem to be any published studies on this topic. As this book goes to press, there is news that Procter and Gamble is marketing the fat substitute under the tradename Olean. Also, the company is planning to use the material in preparing new versions of Pringles fabricated potato crisps.

Use of the liquid forms of the polyol polyesters, such as Olestra, as food can lead to leakage of polyesters from the digestive tract, such as often occurs when mineral oil is taken for medicinal purposes. This is obviously not a desirable feature. It does not occur if the polyol polyester is solid at body temperature, but these higher melting versions taste waxy. Certain intermediate melting polyol esters provide insignificant passive oil loss while at the same time reducing waxiness in the mouth. They exhibit a unique rheology at body temperatures as a result of the formation of a matrix of about 12% solidified molecular aggregations that bind a larger liquid portion. Young et al. (1992) described a method for using such nondigestible materials in combination with regular fats to produce reduced calorie foods such as potato chips, as by frying or spraying.

The scientists at Procter and Gamble have made proposals and filed for patents on means for combatting some of the negative features of Olestra, such as leakage of liquid material from the digestive tract and interference with the absorption of certain vitamins. In the patent of El-Nokaly et al. (1995), it is claimed that adding hydrophobic silica to the ingredient will control or prevent passive oil loss. Others have recommended the addition of some fat soluble vitamins to the ingredient to offset the amount expected to be taken up by the sucrose polyester as it passese through the intestines.

Diol esters — The fat mimetics called cyclohexyl diol esters are the fatty acid diesters of cyclohexanediol, cyclohexenediol, and cyclohexdienediol, and their dimethanol and diethanol counterparts. They are said to be edible and partially digestible (Klemann et al. 1991). At least one of the cyclohexyl diole esters has been said by the inventors to be useful in frying foods such as snack chips and potatoes. Two formulas given by the inventors as examples of possible applications of these materials are shown below.

Butter cream icing
　　Cream the following:
227.0 parts sugar

70.8 parts of a cyclohexyl diester described in the patent
28.4 parts water
14.0 parts nonfat dry milk
 1.4 parts emulsifier
 1.0 parts salt
 1.0 parts vanilla.

Pudding
67 parts milk
11 parts sugar
 5 parts starch
 9 parts water
 3 parts flavor
 5 parts of a cyclohexyl diester described in the patent.

A pet food application is also described, in this case being a mixture that can be extruded to give expanded kibbles.
37 parts hominy feed
17 parts 52% meat meal
13 parts wheat shorts
16 parts of a cyclohexyl diester described in the patent
 9.6 parts corn germ meal
 3 parts wheat germ meal
 0.9 parts dried milk
 1.7 parts beet pulp
 0.5 parts fish scrap
 0.5 parts brewers' yeast
 0.5 parts salt
 0.1 parts vitamin and mineral premix

Long-chain diol diesters, a variation on the previously described compounds, have also been suggested as low calorie fat mimetics (Klemann et al. 1991). The specific compounds, as described in the referenced patent, are aliphatic chains of 11 to 30 carbons having two fatty acid esters or two dicarboxylate-extended fatty acid esters as substituents separated by one or two methylene groups and attached to one end of the chain. Some, or perhaps all, of these compounds are partially digestible but yield lower available calories than ordinary fats and food oils, often in the range of 1 to 6 cal/gm.

Other compounds — Another class of fat mimetics includes a series of complex polyol esters elaborated with fatty acid residues, and/or residues of esters or ethers having an acid function (Kleman et al. 1990). These complex esters are partially broken down in the human body to yield at least two types of aliphatic digestion residues that are more hydrophilic than the original complex polyol ester substrate, but the majority by weight of the digestion residues will be non-hydrolyzable by the normal digestive processes, while a minor amount by weight may be susceptible to cleavage by the action of digestive lipase. Of course, the similarity to sucrose polyesters is evident.

The inventors give an example of a butter cream icing prepared by blending 227 parts of sugar, 70.8 parts of a fat mimetic described in the patent, 28.4 parts of water, 14.0 parts of nonfat dry milk, 1.4 parts of emulsifer, 1.0 part of salt, and 1.0 part of vanilla. The ingredients are creamed in a mixer at medium speed. Organoleptic characteristics of the finished product and its calorie count (and percent of fat reduction) are not given, but it is probable that a 90% reduction in calories from fat is achieved.

Low calorie fat mimetics comprising carboxy-carboxylate esters were patented by Klemann and Finley (1989). These compounds include a carbon backbone substituted with $-CO_2R$ and/or $-CH_2CO_2R$ (carboxylate and/or methyl carboxylate) and with $-O_2C-R$ and/or CH_2O_2C-R (carboxy and/or methylcarboxy) functionalities. The preferred carboxy/carboxylate esters are partially, but not completely broken down in the body. A formula for the type of cookie called vanilla wafers is given. Twenty-five parts of one type of the fat mimetics described in the patent are blended with 100 parts flour, 72 parts granulated sugar, 5 parts high fructose corn syrup, one part non-fat dry milk, one part salt, 0.1 part ammonium bicarbonate, one part dried egg yolk, 0.1 part sodium bicarbonate, and 55 parts water. The dough is rolled, wire-cut to 0.25 inch thickness, and baked by known processes.

Food compositions in which the oil component is comprised of certain polyorganosiloxanes were patented by Frye (1995). Also, a European patent application was filed for the use in food compositions of polyorganosiloxanes with at least 15% by weight organic carbon content and a molecular weight of at least 500. Proposed uses include frying oils, mayonnaise, baked goods, cereals, peanut butter, dairy products, and sandwich spreads. Estimates of caloric reductions from use of this fat mimetic include 27% in blueberry muffins, 55% in ice creams, and 59% in pie crusts.

Microparticulate Proteins

These materials are made by subjecting hydrated proteins in an aqueous suspension to high heat and shear, causing them to coagulate into small round particles, fairly uniform in size. Casein, egg albumen, soy proteins, and some other food proteins have been used as the base for these preparations. When concentrated in a cream-like suspension, some of these materials can be used as fat substitutes. In the best of cases, they give a rich, smooth mouthfeel similar to cream or other fat suspensions. Obviously, since they contain no fat and the protein is not present at a very high level, they are much lower in calorie content than cream. They have the added advantage that they provide the consumer with good quality protein. The ingredient in a ready-to-use (hydrated) form contains from 1 to 2 calories per gram.

Microparticulate proteins can replace fat in a variety of food types, including frozen desserts, cheese, cheesecake, salad dressings (including mayonnaise), cakes, pie crusts and custard fillings, some pastries, bread spreads, chip dips, yoghurt, and sour cream. They can also be used in some cooked food applications such as cream soups, cheese sauces, and casseroles. It is said they retain their fat replacement properties when used as ingredients in food products that undergo a wide variety of manufacturing processes, and treatments such as pasteurizing, canning, and ultrahigh temperature sterilization. They cannot replace fat as a frying medium.

Mayonnaise substitutes and pourable salad dressings can be formulated with microparticulate protein using standard manufacturing processes. Such ingredients have also been used in producing margarines and spreads that are said to have excellent mouthfeel, opacity, spreading texture, and clean flavor. Reductions of as much as 67% of the fat and 50% of the calories can be achieved in these items without difficulty.

Other low-fat or fat-free products that have been made with this ingredient include cheese, yoghurt, sour cream, and cream cheese. Good quality ice cream has been marketed using the material.

Modified protein texturizers, which are similar in many respects to the materials discussed above, are made from a mixture of dried egg whites and whey protein concentrate (or skim milk) in a matrix that includes a food grad gum (preferably xanthan gum) as an adjunct. The matrix is formed by adjusting the pH of an aqueous solution to form a precipitate, draining, and boiling. The resulting partially aggregated milk and egg proteins with the adjunct are in a fibrous form that can be homogenized (Erdman 1990 and Singer and Dunn 1990).

chapter six

FAT REPLACERS AND SUBSTITUTES

The category of fat replacers and substitutes (as defined for the purposes of this book) includes those ingredients that provide some of the desirable characteristics of triglycerides of fatty acids without contributing an appreciable amount of calories (or fat) to the finished product, but which are not chemically synthesized lipid-like compounds and are not special types of triglycerides of fatty acids. They are often based on gels, gums, and other hydrocolloids. Many of these ingredients operate by binding water in a gel-like structure that, under favorable conditions, gives consumer the impression they are eating a food tenderized or thickened by fat.

As summarized by Katz (1996), "From the early days of fat replacement, the key was to take out the fat, add water, and tie the water up somehow or other . . . When calorie reduction involves a reduction in both fat and sugar, the amount and types of ingredients used to control water activity — a key requirement — must be approached in several different ways."

Water activity in foods having reduced fat content is often kept at a desirable high level by including certain carbohydrates that bind the water molecules more tightly than, say, sugar. The carbohydrates may require special treatment to optimize their effects; they may be subjected to high intensity mixing or homogenization techniques, or modified by enzyme reactions, to gain higher water-binding values.

Also included in this rather diffuse group of fat replacers that constitutes the subject of this chapter are what might be called "fat-optimizers." Here we include certain emulsifiers (which may be monoglycerides or other lipids) that act synergistically with the small amount of fat that is often present in the other ingredients, or which can be added. As a result, the finished product (as well as intermediates) appears to have been made with a much larger amount of fat than is actually present. For example, an ingredient tradenamed N-flate has been proposed as an aid in producing shortening-free cakes. This product, which is essentially a blend of emulsifiers, modified food starches, and guar gum on a nonfat milk base, is said to aid in the incorporation of a large number of air cells during batter preparation, so that a cake with fine uniform cell structure and good volume can be made with no shortening.

Many formulas are included in the following text. Although most of them have been provided or published by manufacturers of the fat replacers used in them, it is believed that they have been developed by reputable food technologists and will yield acceptable products. How the products will fare in comparisons with full-fat controls is another matter entirely, and depends greatly upon the slimming motivations of the consumer.

FRUIT DERIVATIVES AS WATER-BINDING INGREDIENTS

Use of fruit purees and of materials based on byproducts of fruit processing operations as fat replacing ingredients has become very common. These items have the advantage of being "natural" to the extent that they are not synthesized, and they may provide some auxiliary advantages (as of flavor, color, acidity, or preservative effects) but their use is somewhat limited by unavoidable side effects due to the presence of materials other than hydrocolloids.

In addition to the familiar thick pastes, powders made from low moisture versions of fruit purees have become available. In one version, a mixture of pears, apples, and plum purees can be obtained at a moisture level of less than 3%. Some of the advantages of the powder form include greater control over the water content of the finished mix, easier dispensing and measuring of the ingredient, and less stickiness.

Table 6.1
FIBER, SORBITOL, AND MALIC ACID IN SOME FRUIT PRODUCTS[1]

Ingredient	Total dietary fiber	Sorbitol	Malic acid
	%	%	%
Dried plums	8.1	16.5	1.57
Raisins	4.0	Trace	Trace
Dates	7.0	Trace	0.52
Figs	9.9	Trace	Trace

Plum/prune Purees

Puree of dried plums, is the same general type of material that has been used for hundreds of years as the principal component of lekvar, an ingredient popular in Europe as a filling and topping for baked goods. A variation of this material has been strongly advocated as a water-binder for reduced-fat foods. It consists usually of about 45% dried plums ground with added water and/or corn syrup. Also available are plum juice concentrates and low-moisture (4% water) plums in various particle sizes. Purees, pastes, concentrates, powders, and spray-dried forms are being offered.

A supplier's representative has said, "You can remove all the fat from recipes, then substitute half its weight or volume with dried plum puree."

Other advantages claimed for these ingredients are: (1) Naturally occurring acids inhibit mold development, allowing the deletion of calcium propionate from recipes; (2) The natural dark color allows replacement of caramel color; (3) Provide the same binding characteristics as eggs; (4) Have more sorbitol, a naturally powerful humectant; (5) contain organic acids that enhance flavor in low-sodium products; (6) Contain pectins, sorbitol, and malic acid that may act as replacements for shortening by providing soft texture, flavor, and moist sensation; and (7) Components of dried plum puree act as emulsifiers.

Formulas have been published demonstrating the use of the lekvar type of prune paste in various types of bakery foods (Sanders 1993). Some of these appear below.

Fat-free chewy brownies

Blend the following ingredients at low speed for one minute, then scrape the bowl and mix at high speed for four minutes.

60.0 parts dried plum puree
95.0 parts granulated sugar
30.0 parts high fructose (71%) corn syrup
11.0 parts liquid egg white
40.0 parts dutched cocoa (10% to 12% fat)
 0.15 parts salt

Add the following to part 1 (above) while mixing at low speed for one minute. Scrape the bowl. The mix at medium speed for four minutes.

100.0 parts all-purpose flour
 2.0 parts sodium bicarbonate
50.0 parts water

Spread the final mixture about one-half inch thick in a lightly greased sheet pan. Bake at 365°F for 20 to 26 minutes, or until the middle part of the sheet springs back after a light touch.

The finished product is said to contain less that 0.5 grams fat per ounce; the calories are 90 per ounce; calories from fat, 5%

Reduced-fat carrot cake or muffins

Blend the following ingredients at low speed for 30 seconds. Scrape the bowl. Mix at high speed until smooth and light.

91.0 parts dried plum puree
89.0 parts granulated sugar
15.0 parts egg white
 1.4 parts vanilla extract
 0.7 parts almond extract

Add the following to stage 1 (above) and mix at low speed for 30 seconds. Scrape the bowl. Mix again at low speed for two minutes or until well-distributed.

    100.0 parts all-purpose flour
      1.9 parts baking powder
      2.4 parts sodium bicarbonate
      1.4 parts salt
      1.0 parts cinnamon
     72.0 parts shredded fresh carrots
     64.0 parts crushed pineapple, in juice
     19.8 parts golden (bleached) raisins
     18.0 parts shelled walnuts, diced
     84.0 parts water

    Deposit the batter into oiled sheet pans or muffin cups, filling them about two-thirds full. Bake at 350°F for about 30 to 35 minutes for cake or 25 to 30 minutes for muffins, judging doneness by the presence of a crumb that springs back after a light touch.

Reduced fat bran muffins

    The following formula was based on a standard commercial formula that yields a product containing one gram of fat per ounce of finished product. The experimental product, contains less than one-half gram of fat per ounce, this coming from the flour and bran primarily, and it contains no added fat.

    31.0 parts granulated sugar
     1.0 part salt
    75.0 parts bread flour
    25.0 parts wheat bran
     2.0 parts baking powder
     1.0 parts sodium bicarbonate
    64.0 parts nonfat milk
    50.0 parts water
    34.0 parts liquid egg white
    30.0 parts dried plum puree

    Blueberry puree is another fruit preparation that has been recommended as a structure-forming material in low-fat bakery products. When chocolate and spice cakes were made with the ingredient and shortening greatly reduced or eliminated, darker crust color was obtained but not much difference in crumb color was observed as compared with the conventional formulations. The experimental cakes were lower in fat and calories but not much difference was found in dietary fiber content.

Modified and Combination Fruit Derivatives

"Fruitrim" is a trademark for a fat replacer formed by bonding unrefined fruit juices with dextrins from grains. The manufacturer states that the combination of mono- and disaccharides from fruit juices with medium- and long-chain dextrins of grains creates a "very effective, sweet-tasting fat replacer with natural humectancy and low water activity." The ingredient is a liquid stable at room temperature. It can be supplied in pails, drums, and tanker trucks. Analytical data are as follows:

Moisture, 22%
Protein, 1%
Ash, 0.7%
Fat, 0.25%
Carbohydrates, 76%.
Glucose, 32% to 38%
Fructose, 16% to 20%
Maltose, 12% to 16%
Sucrose, none to 6%
Other carbohydrates, 6% to 12%
Brix, 77% to 79%
Other specifications include:
Specific gravity, 11.8 lbs per fallon
Viscosity, 4,000 cps at 25° C and 78.5 Brix
Color: Amber
Microbial features:
Total plate count, <500 per gram
Yeast, <10 per gram
Mold, <10 per gram

The manufacturer of Fruitrim has sponsored extensive product development work in efforts to encourage incorporation of this line of ingredients in fat-free bakery products. A few of the formulas arising from these efforts are shown on the following pages.

Fat-free Molasses Cookie Using Fruitrim

	%
Fruitrim	27.49
Molasses	18.33
Egg white, fresh	3.69
Wheat flour, all-purpose	46.67
Allspice, ground	2.70
Sodium bicarbonate	0.79
Salt	0.43

Preparation method

Place Fruitrim, molasses, and egg white into mixer bowl. Mix with the wire whip until the material becomes stiff. Fold in flour, allspice, baking soda, and salt, then mix until smooth. Cool mixture to about 65°F. Place pieces of desired size on a parchment-covered baking sheet, and bake at 300°F for about 15 minutes.

Fat-free Cheese Parfait or Cheesecake Filling

	%
Nonfat cottage cheese	22.55
Nonfat cream cheese	22.55
Nonfat fluid milk	5.75
Lemon juice	5.75
Egg white, fresh	16.10
Sugar, granulated	12.84
Fruitrim	4.82
Water	7.24
Gelatin, dry	2.41

Preparation method

Mix the nonfat cottage cheese until it is smooth. Soften the nonfat cream cheese. Add all the milk and lemon juice and 50% of the sugar to the cottage cheese, and blend. Whip separately the egg whites, remaining sugar, and Fruitrim until a soft peak is formed. Add warm water to the gelatin and when the water has been absorbed, add to the egg whites and continue whipping for about 3 minutes. Blend the egg white mixture into the cheese mixture, avoiding excess agitation. Chill to aproximately 42°F or free, Use either as a stand-along parfait, or as the middle layer in s Swiss-style cheese cake.

Fat-free White Cake

	%
Fruitrim	5.61
Water	10.66
Egg white, fresh	27.97
Cake flour, sifted	21.31
Sugar, granulated	27.98
Baking powder, double acting	1.05
Salt	2.76
Lemon juice	2.66

Preparation method

Put Fruitrim, half the water, and all the egg white into a mixing bowl. Whip until slightly stiff. Add flour, sugar, baking powder, salt, lemon juice, and the remaining water and mix until homogeneous. Do not overmix. Deposit 1 to 1.5 inches thick in lightly greased baking pan. Bake at 300° for 30 minutes. Cool on rack. Ice.

Fat-free Cream Cheese Icing

	%
Nofat cream cheese	60.93
Nonfat sour cream	6.93
Powdered sugar	5.57
Vanilla pudding mix, vanilla	2.93
Fruitrim	9.77
Non-fat margarine	13.87

Preparation method

Place all of the ingredients into a mixing bowl and blend at low speed until smooth. Vanilla can be added.

Low-fat Muffin

	%
Fruitrim	8.23
Egg white, fresh	9.48
Butter	3.94
Vanilla extract	1.34
Skim milk, fluid	8.59
Flour, all-purpose	27.90
Baking soda	0.45
Salt	0.45
Sugar, granulated	13.42
Baking powder, double-acting	0.89
Blueberries, frozen (thawed)	25.31

Preparation method

Put Fruitrim, egg whites, melted butter, vanilla, and skim milk into the mixing bowl. Mix until homogeneous. Add flour, baking soda, salt, sugar, and baking powder to the mix and blend until uniform; do not overmix. Fold in the blueberries. Place measured portions into muffin tins that have been lightly sprayed with a release spray. Bake at 300°F for 12 to 15 minutes, depending on the size of the muffins.

Low-fat Carrot Cake

	%
Fruitrim	16.51
Whole egg, fresh	11.00
Cake flour, unsifted	18.70
Sodium bicarbonate	0.55
Sugar, granulated	4.13
Salt	0.16
Cinnamon, ground	0.77
Allspice, ground	0.04
Carrots, raw, grated	30.26
Raisins	8.80
Pineapple chunks, drained	9.08

Preparation method

Put Fruitrim and the eggs into mixer bowl. Add flour, baking soda, sugar, salt, cinnamon, and allspice on top and mix for three minutes at medium speed. Fold in the carrots, raisins, and pineapple. Place on a baking pan to the thickness of one inch. Bake at 350°F for about 30 minutes. Cool the pan and its contents on a rack.

Pound Cake

	%
Butter	4.96
Sugar, granulated	20.28
Fruitrim	12.41
Vanilla extract	0.34
Lemon juice	0.45
Salt	0.45
Mono- and diglycerides	0.27
Cake flour, sifted	32.68
NFDM, not instant	1.58
Baking powder, double-acting	0.68
Water	6.30
Whole egg, fresh	14.64
Egg white, fresh	4.96

Other Fruit Pastes

Raisin pastes have long been known and used in numerous kinds of baked products, sometimes as a moisture-retaining ingredient. Fig paste is a well-accepted ingredient in fillings for bar cookies. Date paste has been used as a "natural"sweetener and for its humectant and flavor properties.

Several other dried fruit pastes and byproducts remaining from fruit juice preparation have been recommended as fat replacers based on their water-binding qualities. The contention in these cases is the same as for the ingredients previously discussed, that is, the added water remaining in a finished product will give a softness and other textural characteristics ordinarily due to shortening, enabling a large reduction (or complete elimination) of the fatty ingredients. Dried fruit pastes have had a considerable popularity for this purpose. One of these,

LOW-FAT FOODS MADE WITH HYDROCOLLOIDS

Natural, modified, and synthetic gums, mucilages, starches, and other types of hydrocolloids have enjoyed considerable popularity as replacements for fat in certain high moisture types of foods. In combination with water, they convey the sensation of lubricity or greasiness, or a reasonable facsimile thereof, and they also cause the thickening and clinging effects so desired in salad dressings and sauces.

Some of the viscosity-modifying materials that have been suggested as being useful ingredients for the fat-replacing function are polydextrose, maltodextrins, tapioca dextrins, potato starch, microcrystalline cellulose, and gums such as alginates, xanthans, carrageenans, and locust bean gum.

When using these materials, it is often found to be difficult to fully hydrate the water-binding additives, and it may be necessary to allow additional time for water to be taken up by the ingredients. Some suppliers recommend that a paste or gel be made before the water-binding material is added to the other ingredients; the premix may be allowed to age so that some retrogradation occurs. In baked goods, the preparation conditions are critical and may require more precision than some manufacturers are able to provide.

Sausage makers routinely use water-binding materials to adjust the texture of the finished product. Extensions of this practice can be used to imitate some of the features normally provided by the rather high fat content of most sausages. Meat emulsions may use carbohydrate, water, and cellulose mixtures or may simply use higher levels of modified starches that are capable of withstanding heat and shear. Amylose derivatives may be added. Best (1996) describes the use of konjac gum, a glucomannan isolated from the konjac tuber, as a water binding and structural agent in imitation meat products. Bill Thomas of FMC is quoted, "We can get the fat content of summer sausage down from 45% to 8% by using konjac particles to simulate the tiny fat particles in the sausage." The USDA has not yet approved this application, however. It is said that konjac gels will maintain their form even when being grilled at 450°F, certainly a useful property. Imitation bacon rinds have been made with the ingredient.

Development of reduced-fat dips and spreads of the dairy type and soft cheeses usually requires the formation of pasteurized emulsions of skim milk plus non-fat solids with or without pectins, gums, or other viscoisty improvers to yield the smooth, soft, cohesive character expected by the consumer. Depending on the limitations of the process being used, more than one homogenizing step may be required, and additional shearing and mixing steps may be used. Very close temperature control is generally required, and processing tanks may require more than a propeller-type of agitator. Filling may require additional engineering as well, because the product may require more setting time.

Starch

Starches from dent corn, waxy corn, high amylose corn, potatoes, rice, tapioca, and wheat have been recommended for use as fat replacers. Each has some advantages for specific applications, but the starch from ordinary field (dent) corn has probably received more actual use than any of the others — price, plentiful supply, and uniformity are favorable for this ingredient.

None of the native starches is a top choice as a fat replacer because their gelatinized forms do not behave very much like lipids. When hydrolyzed, using basically the same principles as required in the production of corn syrup but different conditions, some useful fat replacers can be produced. Typical of these are the maltodextrins, which will be described in detail in the following section.

Other modified starches intended to replace fat are typically adapted to specific products and processes by using various levels of a combination of cross-linking and substitution reactions.

Glicksman (1991) published a number of formulas illustrating the results obtained when certain commercial corn starches and modified starches were used in low-fat or reduced fat formulas. Two of these recipes are reproduced below.

Fat-free creamy Italian dressing
61.1% water
22.0% vinegar
 5.0% sugar
 1.9% salt
 0.9% onion powder
 0.5% garlic powder
 0.4% red bell pepper granules
 0.2% mustard flour
 0.1% white pepper

0.1% oregano
0.3% xanthan gum
7.5% potato starch maltodextrin

Shortening-free yellow cake

28.5% water
21.6% cake flour
18.3% sugar
17.8% whole eggs
7.0% polydextrose
2.0% nonfat milk solids
1.8% baking powder
1.0% mono- and di-glycerides
0.8% modified starch
0.2% guar gum
0.1% xanthan gum
0.16% sodium bicarbonate
0.2% glucono-delta-lactone
0.35% salt
0.16% butter and vanilla flavors.

There seem to be a few minor problems with the above formula. For example, the leavening system is considerably lower than expected. However, it is presented as a starting point for those who wish to experiment with these fat-replacing ingredients.

Other investigators have found that modified food starches of the proper type can perform very well as hydrocolloids in fat-replacing roles in other types of products. The A. E. Staley company developed a series of modified corn starches tradenamed "Stellar." The ingredient is described as being composed of carbohydrate crystallites produced by a controlled acid treatment of corn starch; in form, it is a fine white powder. When an aqueous slurry containing about 20% to 25% of this powder is shear processed under high-pressure homogenization (8,000 psi and above), a smooth, short-textured "creme" results. This white opaque creme consists of a loosely associated network of sub-micron particles in structured water layers, and it is said to have rheological properties similar to shortening. A formula for a cheese-flavored spread containing only 7.5% fat was developed by A. E. Staley (Thayer 1992). In preparation for the main processing step, a "creme" or gel is prepared by heating a mixture of 27% modified starch (Stellar brand) with 73% water. The formula for the aforementioned spread is shown below.

39.20 parts starch gel
29.63 parts water

22.50 parts cheese powder
5.82 parts nonfat dry milk
0.65 parts Mira-thik 468 starch
0.52 parts instant Tender-Jel 419 starch
0.34 parts lactic acid 88%
0.32 parts flour salt
0.26 parts citric acid
0.20 parts dipotassium phosphate
0.16 parts of a 0.01% solution of Red #40
0.15 parts potassium sorbate
0.15 parts of a 5% solution of yellow dye

Waxy (high amylopectin) corn starch of a specific type has properties fitting it to low-fat formulations, as shown in a series of experiments published by Hippleheuser et al. (1995). These workers had the objective of developing a muffin that contained 3% total fat in 55 gram serving and having an organoleptic rating comparable to a commercial bluebeery muffin containing 15.1% total fat.

The corn starch used was from a variety of waxy maize that yielded starch that was 100% amylopectin and has a unique molecular structure with a shorter and denser branch system. In bakery products, this starch is said to provide a silky texture and uniform cell structure and to have a strong ability to retain moisture so as to lengthen the useful life of the bread or other finished food.

In these experiments, a pregelatinized form of the starch was used in order to increase the batter viscosity, to suspend the blueberries, and to stabilize the aeration of the batter. One of the preferred formulas for blueberry muffins found in the Hippleheuser et al. publication is given below.

Formula — Part 1

16.3% sugar
3.62% corn syrup solids
2.5% shortening
0.5% emulsifier
0.40% salt
0.07% lemon oil 2x flavoring

Formula — Part 2

23.4% flour, bleached all-purpose
1.4% baking powder, double-acting
2.0% pregelatinized dull waxy corn starch
0.4% "butter vanilla 16:1"

Formula — Part 3

15.4% water
18.0% skim milk
 9.92% whole eggs
 6.18% blueberries, flour dusted

Method

Part 1 ingredients were creamed and liquid flavors added. The Part 3 milk and water were added to the eggs and mixed until smooth. Part 2 ingredients were screened and sifted, then mixed with Part 1 and with the liquid ingredients. Blueberries were folded into the batter until evenly distributed. Sixty-gram portions were poured into muffin tins and baked at 400°F for about 20 minutes, until light brown.

Results

An evaluating panel preferred the experimental muffin overall to a full-fat commercial formula control. However, the appearance of the control was preferred. Presumably, the goal of 3% total fat was reached, although it is difficult to reconcile this with the 9.92% whole eggs and 2.5% shortening, together with minor amounts of fat from flour and emulsifier. Perhaps 3% added fat is meant.

Maltodextrins

Maltodextrins are cold water soluble, nonsweet carbohydrate polymers made by hydrolyzing food grade starches to give a dextrose equivalent of less than 20. The advantages of these ingredients as fat replacers (when compared to the unreacted starches) include freedom from unwanted side effects such as gel-formation, retrogradation, and opacity. Generally, maltodextrins are easier to disperse in the batch than are unreacted starches. Maltodextrins are of little value as fat mimetics in dry or crisp products such as crackers, pretzels, and certain kinds of cookies and of no value at all as spray fat replacers.

There can be many kinds of maltodextrins. The extent of hydrolysis strongly affects functional properties. The starch used as a raw material, whether corn, potato, wheat, rice, etc., also has a marked effect on properties of the finished products.

A 35% solids solution of maltodextrin is said to be able to replace fat on an equal weight basis in margarines, chip dips, frozen desserts, and confectionery creams (Anon. 1996C). With this replacement, a net reduction of 7.7 calories per gram of ingredient (not finished product) should result.

The principal function of maltodextrins, when they are acting as fat substitutes, is to increase the viscosity of aqueous systems in which they are present at relatively high concentrations. One writer says that salad dressings, sauces, and gravies that traditionally use oil can be reformulated using a 25% solids maltodextrin solution to replace all the oil and a weight-for-weight basis.

A maltodextrin manufactured by enzymatic conversion of potato starch and having a low dextrose equivalent has been offered as a fat replacer. The trade-named "Paselli SA2," is found in the following formula for blueberry muffins. When used as a fat replacer, the ingredient is first hydrated with three parts of water to one part of maltodextrin and heated to between 180°F and 190°F, then allowed to stand for about 24 hours, "at which point it develops a shortening-like consistency."

Maltodextrins currently being produced from corn starch (and perhaps from other starch sources) have been accepted for inclusion in the GRAS list by FDA. Label declarations would be "maltodextrin" for these ingredients derived from corn and potato starch. Rice- and oat-based fat replacers are sometimes made from whole flour rather than separated starch, and they may in such cases be declared as hydrolyzed oat flour or rice flour.

Table 6.2
FORMULAS FOR BLUEBERRY MUFFINS[1]

Ingredient	Control	Experimental
Flour, all-purpose	26.6	26.6
Blueberries	20.0	20.0
Whole milk, liquid	18.6	18.6
Sugar, granulated	13.2	13.2
Creamtex shortening[2]	9.9	2.5
Whole eggs, fresh	9.9	9.9
Paselli SA2[3]	--	1.9
Baking powder	1.4	1.4
Salt	0.4	0.4
Water	--	5.5

[1] Amounts are stated on a batch percent basis.
[2] Shortening brand of Durkee.
[3] A potato maltodextrin from Avebe America, Inc.
Source: Attributed to Ted Tolvanen (1992)

Preparation method

Hydrate the maltodextrin with the water, and heat to 185°F, then allow to stand for about 24 hours. The dry ingredients are mixed, and creamed with the shortening, then the hydrated maltodextrin is mixed in, followed by the whole eggs and whole milk. Finally, the blueberries are added. After a brief mix to distribute the blueberries uniformly without breaking them up, the batter is deposited in muffin cups in an amount that will depend on cup size and shape.

Results

Appearance and other organoleptic properties of the two batches were similar. The control had 290 calories per 100 gram of baked product, the experimental 230 calories, a reduction of 21%. In the control, 36% of the calories were from fat, in the experimental only 17%.

Polydextrose — Polydextrose is a randomly bonded melt condensation polymer of dextrose, sorbitol, and citric acid. It is not sweet in taste and it is resistant to the action of human digestive enzymes. Microorganisms in the intestines convert the material to various other substances, some of which can be taken up by the digestive system and used as energy sources. The net effect is an energy value of about one calorie per gram.

Polydextrose is used mostly as a bulking agent. It replaces the weight of sugar that is lost when sweetness is due to high potency materials such as saccharin. It functions reasonably well in this capacity in baked goods and confections, which depend on sugar to give them some of the typical structural and textural characteristics desired by the consumer.

Polydextrose does not have the physical properties of fats and oils and cannot be used as a frying medium. In a number of applications, however, the substance exhibits fat-sparing properties — particularly in frozen dairy desserts and baked goods. That is, a product containing polydextrose exhibits mouthfeel characteristic of a higher fat level than is present.

Press reports indicate that polydextrose has been approved for food use as a bulking agent, formulation aid, humectant, and texturizer in the US and several other countries.

Cellulose Derivatives

Among the types or forms of cellulose that have been used for, or suggested for use as, fat replacers in foods are:

(1) Cellulose gel (microcrystalline cellulose)

(2) Chemically modified celluloses, such as carboxymethylcellulose and hydroxypropyl methylcellulose.

(3) Powdered cellulose.

The cellulose used as a food ingredient is based on a highly purified material that can be derived from a wide variety of plant sources, such as wood and sugar beet residue. The purified cellulose may be mechanically ground (to produce powdered cellulose), it can be subjected to chemical depolymerization and wet chemical disintegration (to yield microcrystalline cellulose, etc.), or it can be chemically derivatized (to give carboxymethyl-cellulose, etc.).

A patent describing the use of modified microcrystalline cellulose as a viscosity-increasing agent in no-fat salad dressings was granted to Baer et al. (1991). The method includes the steps of heating and repeatedly (at least twice) shearing an aqueous dispersion consisting of 3 to 10 weight percent of microcrystalline cellulose in water. This micronizing or homogenizing occurs in a high shear zone having a pressure drop of at least 12,000 psi and the shearing is sufficient to fragment the microcrystalline cellulose fragments. The microcrystalline cellulose fragments are reagglomerated under high shear conditions to produce an aqueous dispersion of porous micro-reticulated particles having a mean particle size in the range of 5 to 20 microns and with about 75 percent (by weight) of the particles having a maximum dimension of less than 25 microns. The inventor describes the material as microreticulated microcrystalline cellulose (MMC).

The MMC dispersion is combined with about 20 to 33 percent of xanthan gum, and is then blended with additional food components to provide a low fat or fat-free food product. The food product will have from 0.25 to 4 percent by weight of dispersed MMC, from 50 to 99 percent water, from 1 to 35 percent of carbohydrates, up to 10 percent protein, and less than 7 percent triglycerides.

A fat-free thousand island dressing prepared according to the teachings of the Baer et al. patent was formulated and processed in the following manner. Ingredients are stated on an as is basis.

Formula

43.9519 Water
15.0000 Corn syrup, 25 DE
14.0000 Sugar
  6.5000 Vinegar, 120 grain
  5.5000 Tomato paste
  5.0000 Pickle relish
  2.5000 MMC solids
  2.3000 Partially hydrogenated soybean oil
  1.7500 Salt
  0.4000 Xanthan gum
  0.4410 Stabilizers and acidifiers
  2.6571 Flavors, spices, and colorants

Procedure

The MMC dispersion is placed in a high shear vortex mixer of the Breddo pump design. The xanthan gum and sugar are blended together and then slowly added to the mixer and brought to homogeneity by several minutes agitation. The corn syrup and the other ingredients are next added to the blend under vortex shear conditions. The partially hydrogenated soybean oil is melted and added last so as to evenly disperse it without emulsifying it.

Results

The thousand island dressing is substantially fat-free and yet has a well-rounded, creamy, fat-like mouthfeel.

A simulated chocolate coating containing cellulose gel and carrageenan was reported by Izzo et al. (1995). All true chocolate coatings contain very little water, typically about 1%, and the systems change markedly in eating texture with relatively small changes in fat content. Attempting to reduce the calorie content of these materials of high caloric density by adding water, reducing fat, or other manipulations usually leads to unacceptable end products.

The coatings developed by Izzo et al. range in fat content from no fat to 3% fat. Cellulose gel (tradename Novagel) is used at a 2% level, and the formula specifies 70% sucrose, 0 to 15% other sugars, 14% water, 8.5% glycerine, 0.4% salt, and 4% to 6% chocolate liquor or cocoa powder. The cellulose gel is first dispersed into water with a high shear blender until the mixture "reaches the consistency of sour cream." Other liquids, if any, are incorporated by blending with a low shear paddle mixer to yield a pumpable liquid, then the solid components (except for cocoa) are added. The mixture is heated and mixed for 30 minutes to dissolve sugars and achieve maximum viscosity. Cocoa is added last.

The products of Izzo et al. function well as enrobing materials, and do not require tempering steps, as do real chocolate coatings. The authors state, " . . . several [desirable] functionalities of cellulose gel in low-fat coatings were observed: (1) decreased set time; (2) sugar bloom stability; (3) ability to mimic chocolate rheology; (4) quick flavor release; (5 improved texture." If these coatings can be used in commercial production and meet consumers' expectations for texture, appearance, and flavor they will represent a significant step forward in the availability of reduced fat, reduced calorie coatings. Although the authors did not give an estimate of the calorie reduction as compared to a semi-sweet chocolate coating, it would appear to be at least 30%.

The patent of Izzo et al. also described fat-free and reduced-fat caramels made with carrageenan and the proprietary product Novagel

RCN-15. Full-fat caramel candy generally contains 10% to 11% fat. If the fat content is significantly reduced, the product becomes harder, stickier, and more adhesive. The cellulose gel functions to interrupt the protein and sugar system to yield a shorter, softer, and creamier texture that closely resembles these characteristics of a full-fat caramel.

Pectin

A proprietary specialty pectin derived from citrus peel and standardized by addition of sucrose, has been offered by Hercules as a fat replacer under the trade name "Slendid." The additive forms a gel when mixed with about 20 times its weight of water followed by adding a small amount of a gelling agent such as calcium chloride. The pourable gel is sheared in a standard dairy homogenizer at 500 psig to 2,000 psig and is then ready for adding to the product formula. The gel particles are soft and deformable like fat globules and "can mimic the physical and sensory properties of emulsified fat" (Worthy 1991).

FOODS MADE WITH SPECIALIZED EMULSIFIERS

The formulas of most food products were not developed with a view to minimizing fat content. Generally, enough shortening is included to give the desired volume, appearance, and texture in the finished product. There is usually an opportunity to reduce the fat and yet retain most of the desired features. Emulsifiers assist in such programs, making the most of the decreased amount of fat by making its dispersion more uniform and causing the droplets to be reduced in size.

It is important to specify the right kind of fat. Melting point, or, more precisely, the solid fat indexes, have a definite correlation with the effect of a fat on food quality. If obtaining lubricity is the major requirement, an oil will be more effective than a hard fat. If strength of the dough or viscosity of the batter is a requirement, then a fat with a relatively high melting point would be a more logical choice.

Certain types of emulsifiers, either alone or in combination with gelling agents, have been found useful in making it possible to produce high-moisture low-fat foods. Although these systems have limitations, they perform satisfactorily for several types of formulas ordinarily requiring high levels of fat, and they often permit lowering of calorie count to a very useful extent. Emulsifiers commonly used in food applications are esters of edible fatty acids and polyols such as glycerol, propylene glycol, and sorbitol. These esters can be further modified by esterification with organic acids or by derivatization with ethylene oxide. The emulsifiers used in fluid cake shortenings are usually the alpha gel film formers such as propylene glycol monoester and lactylated monoglycerides — these compounds can be used without oil but are high melting compounds and so must be distributed by adopting high intensity mixing procedures, homogenization, or some other equivalent method.

Some of the emulsifiers used in food systems have the ability to complex with starch, to interact with protein, and to modify the crystalline characteristics of fats in the milieu. These non-emulsifier qualities (which are influenced by the size and shape of the emulsifier molecule as well as by its hydrophilic/lipophilic character) can be more important in a reduced-fat system than the molecule's ability to reduce surface tension and stabilize emulsions.

Emulsifiers in Bakery Products

In formulating bakery products, it has long been noticed that some emulsifiers help to produce differences in texture of the finished product. Saturated monoglycerides tend to cause a soft, resilient crumb, while adding unsaturated monoglycerides leads to a soft, but less resilient crumb. Stearoyl lactylates are also good crumb softeners. A softening effect is one of the attributes of shortenings, and it was found that a relatively small amount of emulsifiers could, sometimes, be as influential as much larger amounts of fats and oils. In this application, then, emulsifiers could function as fat replacers even though most emulsifiers have about the same caloric content, gram for gram, as fats. The weight of fat that is deleted from the formula in these cases can be made up with water or carbohydrate, at a significant net saving in calories per portion.

A few emulsifiers strengthen doughs, making them more elastic and cohesive. This phenomenon is probably due to increasing the access of water molecules to certain functional groups on the protein molecules, Stronger doughs react differently in processing, often causing problems in dividing and molding, and leading to unwanted changes in texture of the finished products. These effects run counter to the desired tenderizing and shortening effects of fat ingredients, and may rule out the use as fat replacers of emulsifiers acting in this manner.

One of the most important characteristics of emulsifiers is the ability to promote and stabilize foams. This is critical in fat-containing cakes, frostings, and aqueous emulsions such as dessert toppings. Normal batters, with high amounts of shortening, are aerated during mixing by the air bubbles that are entrapped in the fat phase. If emulsifiers are present, they aid the dispersion of fat, and therefore, therefore the dispersion of the air bubbles, so the grain of the baked product becomes smaller, and the eating texture (usually) improves, approximately the same effect that would come about as the result of adding more shortening.

Blueberry muffins seem to be a favorite test food for developing dietetic bakery products, perhaps because the flavor and texture contributions of the blueberries conceal some of the other problems that might appear. The formula in Table 6.3, comparing the traditional and fat-free formulas for blueberry muffins, illustrates the general type of substitutions expected in such transformations when emulsifiers are relied upon for textural assistance (Bakal 1994). It is noteworthy that the fat-free formula includes a large amount of egg whites, so that it is technically not possible to describe the product as "egg-free," but it does appear to be permissible to call it "cholesterol-free."

Table 6.3
FAT-FREE BLUEBERRY MUFFINS[1]

Ingredient	Traditional	Fat-free
Sugar	13.0	17.0
Shortening	10.0	--
Whole eggs	10.0	--
Skim milk	18.0	26.0
Flour, all-purpose	26.8	21.0
Baking powder	1.4	1.4
Blueberries	20.0	20.0
Liquid egg whites	--	9.9
Emulsifiers	--	0.8
Xanthan gum	--	0.1
Oat fiber	--	3.0
Salt & other flavors	0.8	0.8

[1] Amounts are stated on a batch percent basis.
Source: Bakal (1994).

A reduced fat ready-to-spread frosting was patented by Glass et al. (1992). The invention is characterized by reduced levels of a defined triglyceride high solids index shortening, high levels of certain emulsifiers, less than 1% of a high strength gelling agent, and elevated moisture levels. The frosting composition has a density of about 0.95 to 1.20 g/cc, and contains less than 6% total fat. A formula for chocolate reduced fat ready-to-spread frosting follows.

Formula

67.58% powdered sugar (12X)
17.50% water
 2.50% corn syrup, 42 DE
 3.33% emulsified shortening (contains about 7% monoglyceride)
 3.00% emulsifier, tradename Bealite 3550 (mainly monoglycerides)
 0.20% emulsifier, blended ethoxylated monoglycerides
 0.33% agar
 5.00% dutched cocoa
 0.13% potassiium sorbate
 0.25% salt
 0.10% buffer
 0.08% citric acid

Process

Mix the agar and water at about 185°F. Add corn syrup and ethoxylated monoglycerides to this dispersion, along with ingredients in quantities less than 1%. Subject the mixture to mild agitation until a uniform blend is obtained. The emulsifier powder, sucrose, and cocoa were added. Finally, the emulsified shortening was added and the mixture was agitated. During the mixing procedure, conducted at about 99°F, air was forced through the mixture. The finished material was cooled to about 99°F, and packaged.

Results

Density was about 1.13 g/cc. Fat content was about 5.6%. Calorie content was about 3.5 cal/g.

From the same source (Van Den Bergh) comes the following formula for a "94% fat free vanilla icing" using an alginate and an emulsifier in combination with small amounts of fat to produce a creamy style of cake decorating or filling material.

Formula

72.70% powdered (12X) sugar
 7.00% nonfat dry milk
 0.10% alginate
 0.08% potassium sorbate
 4.27% shortening ("Creamtex")
 0.67% emulsifier ("Dur-em 204")
 1.00% high performance fat ("Durlite F")
10.88% water at 20°C to 30°C
 3.00% corn syrup solids, 42DE
 0.30% flavoring materials (vanilla, butter, etc.)

Procedure

1. Combine corn syrup solids and water, then add liquid flavorings and mix.

2. Mix powdered sugar, nonfat fry milk, alginate, and potassium sorbate with a whip at low speed until well blended.

3. Slowly add the shortening and emulsifier to the dry mixture and blend on low speed until uniform. While mixing at low speed, add the liquid premix and the high performance fat that has been melted in a separate container.

4. After all ingredients have been added, mix on medium speed for 30 seconds. Scrape down the bowl and mix again for 30 seconds.

Results

According the manufacturer, the above formula gives a product that has 76% less fat and 20% fewer calories than a standard vanilla icing with similar properties.

Emulsifiers in Dairy Type Products

The Van Den Bergh company has published many formulas for reduced fat and no-fat products that incorporate one or more of their long list of proprietary specialized emulsifiers or stabilizers. Typical of these is the following recipe for a reduced fat version of a dairy-based vegetable and chip dip.

Formula — Base mix

90.77% fluid skim milk
2.60% "Duromel" high performance fat
2.00% "Instant Pure Flo" modified starch from National Starch Co.
1.5% nonfat dry milk
1.3% sugar
0.4% "Dur-Em 207" emulsifier
0.83% Flanogen LA, a gum from Sanofi
0.50% salt
0.10% potassium sorbate.

Finished — formula

98.02% of the above mix
1.70% dehydrated vegetables
0.13% garlic powder
0.15% onion powder

Process

1. To skim milk, add salt, potassium sorbate, Duromel, Dur-Em 207, and nonfat dry milk while mixing under high shear and heating to 155°F.

2. Homogenize the mix at 2500 psi through a two-stage homogenizer.

3. Preblend the starch, sugar, and gum. Add this mixture gradually while blending under high shear, meanwhile heating to the 175°F to 180°F range. Hold at finish temperature for 20 seconds.

4. Cool quickly to 70°F to 75°F. Acidify to pH 5.1 by adding Stabilac 112 (from Blanke Baer). Avoid excessive air incorporation. Then, add vegetable mix and seasonings.

5. Pack in selected containers. Store under refrigeration.

Results

According to the Van Den Bergh figures, there is a reduction of about 78% in the fat content of the product, as compared to a control of similar quality, a calorie reduction of about 54%, and a reduction of about 52% in the calories derived from fat.

Maltodextrins in Combination with Emulsifiers

Much experimentation has been done with replacement of fat by combinations of emulsifiers, such as monoglycerides, and with hydrocolloids, such as maltodextrins, and water. Reportedly, considerable success has been obtained.

A commercially oriented type of formula for "Fat reduced Chocolate Pudding," containing maltodextrin, a suitable emulsifier, a gel former, and about 2% added fat (in addition to fat from cocoa) has been published by Van Den Bergh Co. This is reproduced below.

Formula

71,10% skim milk
18.00% sugar, 6X
3.47% maltodextrin, Lodex 5 (Amaizo)
2.50% modified starch, Instant Pure Flo (National)
2.00% high performance fat, Duromel (Van Den Bergh)
1,00% cocoa powder (De Zaan D-11-V)
1.00% cocoa powder (De Zaan D-11-N)
0.30% emulsifier, Durlac 100 W (Van Den Bergh)
0.30% salt
0.30% SeaGel DP 437 (FMC)
0.03% ethyl vanillin

Process

Add salt, cocoa powder, Duromel, and Durlac to skim milk while mixing under high shear and heating to 155°F. Homgogeniz/mix through a two-stage homogenizer (2000 & 500 psig). Blend starch, gum, maltodextrin, and sugar, and add to the milk blend while mixing under high shear and heating to 175°F to 180°F. Hold for 20 seconds. Cool quickly to 70°F to 75°F. Add ethyl vanillin. Pack. Store under refrigeratin.

Results

Although an analysis is not given, it would appear that the total fat content would be less than 3%, this including the ingredient fat, the fat from the cocoa, the emulsifier, and traces of fat in the skim milk.

Sobczynska and Setser (1991) successfully developed a formula and process for devil's food cake in which the shortening was completely replaced by such a combination. The control cake and experimental cake formulas are reproduced below.

Table 6.4
FORMULAS FOR DEVIL'S FOOD LAYER CAKE[1]

Ingredient	Control	Experimental
Cake flour, Sno-Sheen	100.0	100.0
Granulated sugar	121.1	121.1
Hydrogenated shortening	63.2	0
Cocoa, Hershey's	26.27	26.27
Nonfat dry milk	15.74	15.74
Whole eggs, fresh	108.9	108.9
Baking powder, double-acting	6.5	6.5
Sodium bicarbonate	3.27	3.27
Salt	5.7	5.7
Vanilla extract	1.33	1.33
Xanthan gum, Keltrol TF	0.078	0.078
Polydextrose, Pfizer	15.0	15.0
Water, distilled	90.0	90.0
Emulsifier[2]	0	0.82
Potato maltodextrin[3]	0	63.2

[1] Amounts are stated on a flour weight basis (i.e., flour = 100%).
[2] Emulsifier was monoglyceride Myverol 18-06 from Eastman, added at 10% of the weight of the potato maltodextrin (dwb). Gels were formed prior to addition to the mixture by adding to a portion of the water used and adjusting the pH of the dispersion to 7.1.
[3] The potato maltodextrin was Paselli SA2 from Avebe America, Inc.
Source: Sobczynska and Setser (1991)

Preparation method

The preparation procedures were rather complex and involved the pre-forming of gels from emulsifiers and maltodextrins before they were added to the batter, and other manipulations not common to commercial layer cake production. The original reference should be consulted if the reader desires additional details.

Results

Most of the variables yielded cakes of satisfactory appearance and texture. The researchers concluded that total replacement of shortening with combinations of maltodextrin and emulsifiers was feasible. In their series of tests, the researchers also tested sucrose esters and sorbitan monostearate with good results.

Other Types of Emulsions

A group of ingredients offered under the tradename "Veri-Lo," have the form of fluid emulsions. They are referred to as fat extenders because they contain a small amount of lipids. The manufacturer (Pfizer, now Cultor Foods) describes them as micronized fat particles imbedded on the surface of deformable particles composed of aqueous gel, and says the fat-rich surface delivers the taste and mouthfeel characteristics of fat and oil droplets. These emulsions are versatile and can be produced with a number of different fats depending on the requirements of the food application. Some of the recommended applications include mayonnaise, Italian salad dressings, creamy salad dressings, sauces, and gravies. Dairy flavored versions are available to improve the cream simulation mode.

chapter seven

HEALTH EFFECTS OF CERTAIN LIPIDS

Diet modification by reducing the consumption of total lipids has been discussed in preceding chapters. The current chapter deals with specific effects on health of different varieties of lipids, for example, polyunsaturated fats and cholesterol.

FAT STRUCTURE AS RELATED TO NUTRITIONAL CONCERNS

All common fats and edible oils are triglycerides in which three fatty acids have combined with a single glycerol molecule, the acid groups of the fatty acids chemically combining with the hydroxyl groups on the glycerol molecule, leading to the production of fats and water. This process can be reversed, hydrolyzing the fats with alkalies to enzymes to yield fatty acids and glycerol. Plants and animals have very efficient enzyme systems for producing and rearranging fats and oils, and also for utilizing fats and oils both as energy sources and as building blocks for more complex systems.

Differences in Fats

A natural fat or oil will ordinarily contain several different fatty acid residues, and a single molecule may contain two or three different fatty acids. Since the natural products are a mixture of different molecules, they do not have sharp melting and solidification points as do pure substances, i.e., those made up of only one kind of molecule.

The wide variation in melting points and other physical properties of fats are caused by differences in the fatty acids; chemical and physiological responses may also be affected by these differences.

The physical status of fats in shortenings can be very important to their processing response and general consumer acceptability, and some of these changes are made possible by the methods that change the micro-crystalline or quasi-crystalline structure of the material without causing detectable alterations in the chemical structure of the triglycerides. It is believed that these purely physical changes are of little or no importance relative to the body's reaction to the products.

Fats are characterized by their fatty acids, glycerol being a simple molecule having no stereoisomers or other variables causing complications in its analysis or characterization. There are several modifications of fatty acid molecules, however.

It is not necessary that all three fatty acids on a glycerol molecule be of the same type, in fact, it is not unusual to find two or three different fatty acids in one molecule of fat.

Glycerol molecules having only one or two fatty acids attached to them are fairly common, and are called monoglycerides and diglycerides, respectively. These compounds are lipids but are not considered to be fats or oils. Many useful food emulsifiers are based on monoglycerides.

Fatty acids all consist of a chain of carbon atoms, most of which have two hydrogen atoms attached to them by chemical linkages, but with the carbon atom at one end of the molecule having three hydrogen atoms bonded to it, and the carbon atom at the other end having an oxygen atom and a hydroxyl group attached. All other carbon atoms have either one or two hydrogen atoms (usually two) linked to them. If there is only one hydrogen atom on a carbon atom, this carbon will be found to be linked to one of its neighbors by a so-called double bond.

These double bonds cause modifications in the chemical, physical, and physiological behavior of fatty acids containing them. There can be more than one double bond in a fatty acid, and these polyunsaturated fatty acids (PUFAs), and the fats and oils containing them, have received a great amount of attention from nutritionists.

Formic acid, acetic acid, and propionic acid (with one, two, and three carbon atoms, respectively) might be called fatty acids since they meet the chemical definition, but it is customary to consider butyric acid, with four carbon atoms, as the smallest typical fatty acid, especially when discussing food components.

Chemistry of Fatty Acids

Fatty acids are differentiated by chain length (number of carbon atoms), in the number of double bonds that are present and their location in the chain, and in geometric positioning of hydrogen atoms. The carbon atoms in a fatty acid molecule or residue are bonded together in a straight chain; if there are any branched chains in natural fats, they must be very rare. It has been said that in excess of 90% of the naturally occurring fatty acids found in foods consumed in a typical American diet have an even number of carbon atoms.

A dictionary definition of fatty acids says they are of the series of saturated acids having the general formula $C_nH_{2n}O_2$, as formic acid, acetic acid, etc. — so-called because some of the members, such as stearic and palmitic acids, occur (in their glyceryl esters) in the natural fats, and are fat-like substances. Although formic acid, acetic acid and propionic acid (one, two, and three carbons, respectively) have the same general chemical structure as fatty acids, they are not generally considered as such, at least not in dietary discussions. Most tables of fatty acids will begin with butyric, the fatty acid containing four carbon atoms and run up to at least 24 carbon atoms. Contrary to the definition, current usage refers to both saturated and unsaturated compounds of this general type as fatty acids. Indeed, many unsaturated "fatty acids" are found in natural fats and are of great interest and importance to nutritionists, processors of fats and oils, and marketing personnel.

Table 7.1
COMMON SATURATED AND MONOUNSATURATED FATTY ACIDS

Common name	Standard chemical name	saturated: unsaturated bonds[1]
Saturated Fatty Acids		
Butyric	Butanoic	4:0
Caproic	Hexanoic	6:0
Caprylic	Octanoic	8:0
Capric	Decanoic	10:0
Lauric	Dodecanoic	12:0
Myristic	Tetradecanoic	14:0
Palmitic	Hexadecanoic	16:0
Stearic	Octadecanoic	18:0
Arachidic	Eicosanoic	20:0
Behenic	Docosanoic	22:0
Lignoceric	Tetracoanoic	24:0
Monounsaturated Fatty Acids		
Myristoleic	*cis*-9-tetradecenoic	14:1(9)
Palmitoleic	*cis*-9-hexacecenoic	16:1(9)
Oleic	*cis*-9-octadecenoic	18:1(9)
Vaccenic	*cis*- and *trans* octadecenoic	18:1 cort(11)
Gadoleic	*cis*-9-eicosenoic	20:1(9)
Erucic	*cis*-13-docosenoic	22:1(13)

[1] Number in parentheses indicates position of unsaturated bond.
Source USDA Handbook 8 and Sullivan (1994).

In saturated fatty acids, there are only single bonds between adjacent carbon atoms. A large part of the saturated fats consumed by Americans comes from the stearic and palmitic acids found in beef and other animal fats, but the hydrogenated vegetable oils found in the cooking shortenings so widely used in homes and industry are also major sources of saturated fats. Unsaturated fatty acids contain one or more double bonds, which may theoretically be located anywhere along the chain. The number of unsaturated bonds and their position along the chain affect the physical, chemical, and nutritional properties of fatty acids.

Monounsaturated fatty acids have one double bond between two adjacent carbon atoms in a chain. The predominant monounsaturated fatty acid in foods is oleic acid; it is present in most common fats and food oils. Other fairly common monounsaturated fatty acids are caproleic, lauroleic, myristoleic, elaidic, and vaccenic (all found in butterfat and perhaps elsewhere), and palmitoleic (found in beef fat and some fish oils).

Polyunsaturated fatty acids (PUFAs) contain two or more double bonds in the chain of carbon atoms. Although less common than either saturated or monounsaturated fatty acids, they are nonetheless widely distributed in nature and are found in numerous foods and food ingredients; some of them, at least, are essential parts of the metabolic processes in plants and animals. Others may constitute merely means of storing energy in a relatively inert form.

Fatty acids with up to six double bonds are known — arachidonic (four double bonds) is found in lard, and some fish oils contain PUFAs with five and six double bonds. Length of the carbon chain, extent of saturation, positional isomerism (place where the double bond occurs in the chain), and geometric isomerism (relative orientation of the hydrogen atoms on the double-bonded carbons) affect the body's utilization of and response to fatty acids and so are of legitimate concern to the designer of foods for dietetic purposes.

Two important PUFAs in the diet are linoleic acid and α-linolenic acid, which are found in many plant oils. Two PUFAs which have been called "omega-3 fatty acids" are eicosapentaenoic acid with 20 carbon atoms, and docosahexaenoic acid with 22 carbons; they are found mostly in marine fish, and have been touted as being helpful in reducing the risk of heart disease in people who consume them regularly.

The major polyunsaturated fatty acid in the American diet is linoleic acid. It accounts for more than 50% of total fatty acids in such vegetable oils as soybean, corn, and cottonseed oil. Table 7.2 lists some of the polyunsaturated fatty acids by their trivial or common name and their standard chemical name.

Table 7.2
SOME POLYUNSATURATED FATTY ACIDS FOUND IN FOODS

Common name	Standard chemical name	Number of double bonds[1]
Linoleic	*cis*-9,*cis*-12-octadecadienoic	2
Linolenic	octadecatrienoic (9,12,15 all cis)	3
Eleostearic	9,11,13-octadecatrienoic	3
gamma-Linolenic	6,9,12-octadecatrienoic	3
Moroctic	4,8,12,15-octadecatetraenoic	4
Arachidonic	5,8,11,14-eicosatetraenoic	4
Timnodonic	4,8,12,15,18-eicosapentaenoic	5
Clupanodonic	4,8,12,15,19-docosapentaenoic	5
Nisinic	4.8,12,15,18,21-tetracosahexaenoic	6

Source USDA Handbook 8 and Sullivan (1994).

Nomenclature

As a step preliminary to discussing the role of special types of fat and fat components in the diet and the ways these materials can be used to improve the nutritional contribution of specific foods, a brief discussion of the nomenclature applied to fats and fatty acids is necessary.

Many fatty acids have a common name, often derived from the name of the organism from which the substance was first isolated, or from plants or animals that contain an abundance of fats that include the fatty acid. Thus, we have arachidic, named after *Arachis*, the genus name of the common peanut (*Arachis hypogaea*). Also, some fatty acids are named after one of their easily discernible qualities, as capric acid, based on the Latin word for goat, relating to the odor of the substance. There is no way of getting insight into the chemical structure from the common name of the fatty acid. The following discussion is restricted to traditional usage in food chemistry.

All fatty acids have a chemical name, based on conventions established by, or recognized by, the International Union of Chemistry. At first glance, these appear to be merely a meaningless jumble of letters and numbers because of their unfamiliarity and complexity, but the situation may not be quite as bad as it at first appears.

Table 7.3
FATTY ACID PROFILES OF COMMON FATS[1]

Oil source	C12:0	C14:0	C16:0	C18:0	C18:1	C18:2	C18:3
Corn	--	--	11.0	2.0	26.0	60.0	1.0
Soybean	--	--	11.0	4.0	24.0	54.0	7.0
Cottonseed	--	1.0	22.0	3.0	19.0	55.0	1.0
Peanut	--	--	11.0	2.0	48.0	32.0	1.0
Canola	--	--	4.0	2.0	62.0	22.0	10.0
Olive	--	--	13.0	3.0	74.0	9.0	1.0
Palm	--	1.0	45.0	4.0	40.0	10.0	--
Coconut	48.0	19.0	9.0	3.0	7.0	2.0	--
Lard	--	2.0	27.0	14.0	46.0	11.0	--
Butter	4.0	12.0	28.0	12.0	30.0	2.0	1.0

[1] The total fatty acid content is taken to be 100% (rounding of figures may lead to totals different than 100%). The heading "C18:3" indicates there are eighteen carbon atoms and three double bonds in the fatty acid, etc. Peanut oil also contains 5% longer chain saturated fatty acids. Coconut oil also contains 13% shorter chain saturated fatty acids. Butter also contains 11% shorter chain fatty acids.
Source: Hall (1989), modified.

The names of all fatty acids end in "oic," a suffix used for naming acids in general, such as "benzoic acid." Immediately preceding this suffix will be either "an" or "en." The former tells you that there are no double bonds in the compound, while the latter indicates that one or more double bonds exist between carbon atoms. Immediately preceding the "en" may be found "di," "tri," "tetra," "penta," or "hexa," which indicates that there are two, three, four, five, or six double bonds; if there is only one, no special term will be found at this point. Immediately preceding the indicator of the number of double bonds (or immediately before "an" or "en," if there is one or no double bond), there will be found a set of letters identifying the total length of the carbon chain, "hexa" for six, "octadec" for eighteen, "eicos" for twenty, etc., all based on Latin numbers because it would be too simple to name an acid "fouranoic" or "eightenoic." Actually, in a more solemn mood, it is recognized that language differences make the use of Latin more reasonable than, say English, since we all know many more people speak Latin than English.

Preceding the word, now complete, we will find one or more numbers, if the acid being named has one or more double bonds.

At the beginning of the names of unsaturated fatty acids may be found the designations *cis* or *trans*, which describe the spatial configurations possible in fatty acid molecules or residues containing one (or more) double bond between carbon atoms, the former indicating the hydrogen atoms on the adjacent double-bonded carbons are on the same side of the molecule and the latter indicating they are on opposite sides of the molecule. The relative rigidity of the double bond prevents rotation of the carbons so that the attached hydrogens cannot freely move back and forth, it being the latter movement that makes the *cis* and *trans* designations meaningless for carbon pairs joined by single bonds.

Nutritional Considerations

The roles played by various kinds of fats and fatty acids in the metabolic processes of normal human bodies are still being elucidated. There are relatively few points on which universal agreement has been reached. Many physiological reactions have been clarified during the past few years, although some of the current beliefs appear to be based on evidence that is less than compelling. The following discussion should be understood as an attempt to present current wisdom without being dogmatic. It is reasonably certain that specific kinds of fatty acids, and the fats containing them, are somehow related to circulatory disorders; the cause and effect chains, however, contain many weak links.

Essential fatty acids — In any organism, many different kinds of fats can be found, but it is common that one, or a few, compounds predominate. The most common varieties are no doubt the forms which have conferred upon the organism a survival advantage as compared to other fats, although the mechanism by which this advantage is secured is often not at all clear. The organism can obtain necessary fats and fatty acids from outside sources or manufacture it from other compounds present in the body. In some cases, fatty acids are needed that cannot be produced by the organism, so they must be obtained from food. These are called "essential fatty acids."

Current nutritional thought has it that linoleic acid and α-linolenic acid are essential dietary factors in humans because they are the only two fatty acids needed by the body for growth and survival that cannot be synthesized by the human metabolic system. After ingestion, the essential fatty acids are changed to arachidonic acid, which has 20 carbon atoms and is the precursor of eicosanoid compounds such as prostaglandins, thromboxanes, and leukotrienes. These are substances profoundly influencing many cellular reactions involved in the functioning of human cardiovascular, respiratory, renal, endocrine, skin, nervous, and immune systems.

Excessive or unbalanced syntheses of eicosanoids can result in pathological conditions such as thrombosis, asthma, and kidney disfunctions. In particular, linoleic acid deficiency is said to cause diminished growth, skin scaling, impaired wound healing, sterility, and increased metabolic rate. Linolenic acid deficiency may affect development of visual and neural functions of infants.

In their obsessional tinkering with the eating habits of the public, certain government and academic nutritionists have not skipped over essential fatty acids (EFAs), even though EFA deficiencies have been seen only in patients with disorders that have affected their fat intake and absorption or who are subsisting on grossly an inadequate (in amount or variety) food supply. The American Heart Association has recommended that up to 10% of a person's daily caloric intake be derived from PUFAs, this being a level thought to take full advantage of the cholesterol-lowering benefits of unsaturated fatty acids. Some experts have suggested a daily intake of 14 grams of linoleic acid and 3 grams of linolenic acids. There is no requirement for linoleic or linolenic in the edition of *Recommended Dietary Allowances* available at the time this article is being written.

, **Unsaturated fats** — The current widespread belief that unsaturated fats are "better for you" has resulted in the development of a sizable market for foods containing fats that have a higher than usual percentage of fatty acids with one, two, or three unsaturated bonds. A commercial disadvantage of supplying this market is the well-known tendency of unsaturated fats to develop oxidative rancidity at rates incompatible with normal distribution patterns; exceptions may be in retail bottled oil and dressings, where access to oxygen is restricted until the container is opened. In the current U.S. retail market place, the largest consumption of edible oils is as margarine, spreads, dressings, retail bottle oils, and frying.

Even the most dedicated artery-watcher will tend to reject foods having a strongly rancid odor. An oil's susceptibility to oxidation tends to be influenced mostly by its content of linolenic (18:3) and linoleic (18:2) acids. Of course, it is easy enough to hydrogenate food oils by well-known and commercially available methods. Hydrogenated vegetable oil shortenings with good stability and excellent functional properties have been manufactured by these methods in large quantities for many years. Unfortunately, this approach leads to the need for labeling disclosures that may have a negative impact on consumer purchases. This conflict between consumers' label preferences and functional problems has been regarded as a marketing opportunity by food manufacturers.

Attempts to remedy the susceptibility to oxidation of unsaturated (and, particularly in polyunsaturated) fats has led to projects having as their goal the development of plants yielding natural fats that are not only high in polyunsaturated fatty acids but also have good resistance to oxidation. The disciplines primarily involved are plant breeding and molecular genetics. The principle candidate, at least currently, for promotion as a low-saturated fatty acid/healthful oil is canola (Erickson and Frey 1994). Other researchers have developed strains of soybeans yielding oil containing relatively low amounts of linolenic acid. This oil is being marketed by Kraft under the tradename Soy·LL. In addition, a new sunflower variety yields an oil that contains about 80% oleic oil with a high level of dietary monounsaturates (Duxbury 1992). The high oleic content provides a longer shelf-life (greater resistance to oxidative rancidity).

Oil is sprayed on several types of cookies and crackers after they are baked in order to give them a glossy appearance, a richer taste, and a better texture. The "Ritz" cracker is a famous variety so treated. These "spray oils" must have high stability since they are spread thinly over a large surface area and so are very much exposed to oxygen. As a result, spray oils were generally specified to be low in unsaturated fatty acids and essentially free of linolenic and linoleic acids. Hydrogenated fats were usually employed, leading to the necessary label declarations that were objectionable to some potential purchasers of the product. The previously quoted reference of Erickson and Frey reported the results of a test in which the storage stability of plain crackers sprayed with low-linolenic-acid soybean oil was compared to crackers sprayed with the usual hydrogenated oil. The samples were stored in dark, constant temperature rooms at 72°F and 95°F for five months. Monthly samples were subjected to sensory evaluations by an expert panel. Generally, comparable results were obtained.

There are many food applications that require a plastic fat ingredient if a normal appearing product is to be obtained. Many baked products fall into this category. The success of creaming, laminating (as in true Danish pastry and in puff pastry), preparing buttercream icings, and many other bakery processing steps depend heavily upon the use of a shortening that contains a large percentage of solid lipids at the temperature of processing. In most cases, this necessitates the use of a hydrogenated shortening. Natural fats that meet this requirement include leaf lard and some tropical oils or their high-melting fractions.

Geometric isomers of fatty acids — The double bond found in unsaturated fatty acids hinders the rotation of the carbon atoms joined by it. Consequently, the fatty acid molecule is said to become more rigid in the vicinity of the double bond and, in particular, the single hydrogen atoms on each of the carbon atoms can be identified as being fixed either on the same side of the molecule, or one on each side of the carbon molecule, "side" in this case being a relative term. Standard practice adds "cis" to the names of those molecules in which the two hydrogen atoms have been shown to be on the same side and "trans" if the hydrogen atoms are on opposite sides;

generally the terms are printed in italics.

This is not strictly an academic consideration; a *cis* fatty acid can have much different physical and physiological properties than the *trans* fatty acid having the same chemical formula. For example, both elaidic acid and oleic acid are made up of 18 carbon atoms, 34 hydrogen atoms, and 2 oxygen

---

Table 7.4

SOME SOURCES OF MAJOR FATTY ACID TYPES

Saturated	Unsaturated Fatty Acids	
	Mono-unsaturated	Polyunsaturated
Medium chain, 6-12	**Omega 9, high oleic**	**Omega 6, high linoleic**
Kernel oils	Olive	Corn
Babassu	Canola	Cottonseed
Coconut	Safflower, hybrid	Soybean
Cohune	Sunflower, hybrid	Safflower, regular
Palm kernel		Sunflower, regular
Tucum		
MCT-oil		**Omega 6, GLA oils**
Long chain, 14-24		Black currant
		Borage
Cocoa butter		Primrose
Dairy fats		
Lard		**Omega 3, high linolenic**
Tallow		
Palm		Linseed
Stearine		Fish oils
		Menhaden
		Salmon
		Mackerel
		Tuna
		Anchovy

Source: Babayan (1989), FDA Handbook 8, Best (1989), and others.

atoms, and they both have a single double bond located at the "9" position. However, oleic acid (the *cis* isomer) melts at 13°C while elaidic acid (the *trans* isomer) melts at 44°C. There are other differences, as well. The properties of *trans* isomers are usually closer to the properties of saturated fatty acids of the same chain length than are the *cis* forms.

It appears that, in general: the longer the chain length, the higher the melting point; the more unsaturated bonds, the lower the melting point for the same chain length molecules; and, *cis* isomers have lower melting points than *trans* forms, other characteristics being the same.

Since interest in this group of compounds has developed only recently, there is not as much information on their quantity in common foods as there is on other types of fatty acids.Most of the unsaturated fatty acids found in nature are of the *cis* geometric isomer type, although *trans* fats are found in milk and butter and in some other natural foods. In commercially produced foods, the most significant source of *trans* fatty acids is hydrogenated shortenings — the catalytic hydrogenation process creates this type of fatty acid, sometimes in relatively large amounts. The *trans* fatty acid content of partially hydrogentated vegetable oils can range between 5% and 45% depending on the properties of the catalyst, hydrogenation conditions, and the iodine value of the vegetable oil.

Average Americans get about 34% to 36% of their calories from fat, about 12% to 14% from saturated fat, and about 2% to 4% from *trans* fats. There is no clear-cut evidence either of special health benefits from, or disadvantages of, *trans* fats, although some researchers have claimed that *trans* fats in partially hydrogenated oils increase LDL cholesterol, compared to non-hydrogenated oils. The general thrust of these arguments appears to be that trans-unsaturated fats have the same bad effects as saturated fats and it is therefore misleading for packagers to include *trans* fats in the total of supposedly beneficial unsaturated fats (Allen 1995C). Some non-profit organizations and lobbying groups have been trying to get the FDA to disallow health claims on labels of products containing significant amounts of such lipids.

It appears, from results of some laboratory studies, that the *trans* conformation affects the ability of enzymes to bond to the molecule and therefore limits digestive and other body reactions involving fatty acids and fats, probably placing the *trans* isomers at an intermediate level between reaction rates of saturated fatty acids of the same chain length and their *cis* counterparts.

Recent publications have suggested that trans fatty acids might interfere with the biosynthesis of long-chain polyunsaturated fatty acids in infants and thereby retard their early growth and development. As a result, one researcher has advised that "it appears prudent to keep the levels of trans fatty acids in infant formulas as low as possible" (Sullivan 1994).

Table 7.5
ALLEGED SPECIAL FUNCTIONS OF CERTAIN FATTY ACIDS

Type of fatty acid	Special function or effect
Saturated	
MCTS[1]	Rapid source of energy.
Lauric (12:0)	Hyperlipidemic; hypercholesterolemic; prothrombotic.
Myristic (14:0)	
Palmitic (16:0)	
Stearic (18:0)	"Neutral" or hypolipidemic; precursor of oleic acid
Monounsaturated	
Oleic (18-1n-9)	Hypolipidemic/hypocholesterolemic; precursor of eicosatrienoic acid (20:3n) in essential fatty acid insufficiency.
Elaidic (18:1 *trans*)	Analogous to 18:0.
Erucic (22:1n-9)	Impaired FA acid oxidation in rat heart.
n-6 polyunsaturated	
Linoleic (18:2n-6))	Essential fatty acid (45 mg/kg/day). Component of acylglucocereamides. Precursor of arachidonic acid. Hypolipidemic relative to saturated FA. May be hypotensive. Increases membrane fluidity.
ϒ-linolenic (18:3n-6)	Precursor of eicosatrienoic acid and arachidonic acid
ϒ-homolinolenic (20:3n-6)	Precursor of PGE$_1$ series of eicosanoids.
Arachidonic (20:4n-6)	Membrane fluidity; eiconsanoid precursor.
n-3 polyunsaturated	
α-linolenic (18:3n-3)	Hypolipidemia; membrane fluidity; precursor of EPA and DHA (essential?); reduces eicosanoid synthesis.
Eicosapentaenoic (20:5n-3)	Hypolipidemic; reduces arachidonic acid synthesis and eicosanoids; precursor of PGI$_3$, TXA$_3$; precursor of TXB$_5$.
Docosahexaenoic (22:6n-3)	Hypolipidemic; essential for vision, neural membranes?; reduces arachidonic acid synthesis; reduces eicosanoid in some cells (macrophages).

[1] A mixture of medium-chain triglycerides, primarily caprylic and caproic.
Source: Kinsella (1988), as modified and expanded.

Ingestion of TFAs has also been linked to decreased amounts of high density lipoproteins in the blood, increased concentrations of low density lipoprotein-cholesterol fractions, and, by extrapolation, to increased risk of cardiovascular disease.

Omega-3 fats — Certain European food manufacturers and ingredient suppliers have been trying for well over two decades to find a niche for omega-3 fats. These are fats, found in fish and perhaps elsewhere, containing polyunsaturated fatty acids that have the first double bond positioned after the sixth carbon atom from the methyl (CH_3 end of the chain). The fatty acids in which this occurs are eicosapentenoic and docosahexaenoic. They are evidently precursors of a wide range of metabolic regulators (eicosanoids) that inhibit blood clotting, minimize inflammatory and immune responses to tissue injury, lower blood pressure through vascular dilation, and protect blood vessel walls from pro-oxidants.

It is suggested that introducing such materials into the diet may reduce the chance of heart disease, and it is also said they affect favorably the development of the brain, spinal cord, and retinas of human fetuses. The ideal level of daily intake of omega-3 fats is said by the British Nutrition Foundation to be one to three grams per day.

Fish oils have been staple items of commerce for many years in Europe, but nearly all of the supply is processed by hydrogenation, which eliminates the omega-3 unsaturation on which all of the current speculation is based. Therefore, any attempt to correlate health status and mortality statistics with the consumption of omega-3 based on disappearance of fish oil in Europe has no logical foundation.

According to proponents, the benefits of omega-3 oils lie not in their effects on blood cholesterol, but in their preventive effects on plaque and blood clot formation (Best 1989). It has been said that omega-3 fatty acids protect laboratory animals against cardiovascular disease even under conditions of elevated LDL blood cholesterol levels.

In the foods category, the fish oils which are found to contain significant amounts of omega-3 fatty acids have not been very popular because of their organoleptic characteristics; also, they appear to be more expensive and less versatile than, for example, soybean oil or palm oil. In recent news reports we learn that low-fat spreads containing relatively high levels of omega-3 fatty acids have been introduced in Ireland and Denmark, while several breads enriched with the material have entered the market in Scandinavian countries.

It remains to be seen whether or not omega-3 will become a fad in the US. Formulating bakery products in which enough fish oil derivatives have been used to replace ordinary shortening should not be difficult; it is said that current versions of omega-3 supplements do not have a fishy taste or aroma.

Getting FDA clearance to the making of health claims on food labels was virtually impossible in former times. It is still difficult. The FDA states that it will issue regulations authorizing a health claim only when it determines, based on the totality of publicly available scientific evidence (including evidence from well-designed studies conducted in a manner that is consistent with generally recognized scientific procedures and principles), that there is significant scientific agreement, among experts qualified by scientific training and experience to evaluate such claims, that the claim is supported by such evidence. There are many companies and nonprofit organizations working to get the government to loosen such restrictions and also to provide guaranteed exclusivity for a period of years to companies that conduct studies justifying such claims.

Food enrichment ingredients suitable for increasing the content of polyunsaturated acids, and especially of the omega-3 types, are being offered by firms such as Hoffman-La Roche. Their trade name for these ingredients is "ROPUFA," and there are evidently several of them. A recipe for mayonnaise made with one of these additives is shown below.

71.25% sunflower oil
3.75% ROPUFA '30' n-3 EPA oil
15.00% egg yolk
7.00% mustard
1.00% salt
2.00% lemon juice

The ROPA '30' n-3 EPA oil contains a minimum of 30% of n-3 polyunsaturated fatty acids.

A formula and procedure for yeast-leavened bread has also been developed by the supplier.

Formula

1400.0 g semi-white wheat flour
500.0 g rye flour, dark
750.0 g milk, fresh
750.0 g water
50.0 g salt
75.0 g leavening
31.0 g ROPUFA '10' n-3 powder

Method

The dough is kneaded for ten minutes in a spiral mixer. Dough temperature is to be 28°C. Ferment for 30 minutes. Scale at 400 grams. Bake in zones of 250°C to 220°C. Finished loaf weight should be about 335 g.

Results

The total of eicosapentaenoic and docosahexaenoic fatty acids (EPA+DHA) in the bread is 1,000 ppm.

A chocolate flavored milk drink has also been developed with polyunsaturated fatty acid supplementation. The formula follows.

Formula

93.98% milk
  0.1% ROPUFA '30' n-3 EPA oil
  0.8% cocoa powder D21 (Dezaan)
  0.02% stabilizer
  0.1% milk-chocolate flavoring
  5.0% sugar

Method

Heat the milk to 65°C and add the dry ingredients, the ROPUFA oil and the flavoring. Homogenize at 220 bar. Heat to 135°C to 145°C for 4 to 6 seconds and fill under sterile conditions.

A formula and method for manufacturing a fruit-flavored functional drink that provides in a liter not only 2.50 grams of "ROPUFA-30 n-3 EPA oil" but also 0.15 gram of α-tocopheryl acetate, 0.20 gram of ascorbic acid, and 0.02 gram β-carotene has been developed by scientists at Roche. Details are given below.

Formula for 200 grams base mix

167.00 g orange concentrate at 65° Brix.
  2.50 g ROPUFA '30' n-3 EPA Oil
  0.15 g alpha-tocopheryl acetate
  0.20 g ascorbic acid
  2,00 g beta-carotene 10% CWS as a 10% solution
  0.20 g sodium benzoate
 10.00 g pectin solution 5%
 17.75 g water
  0.20 g orange flavor

Preparation Method

Add the pectin solution, beta-carotene, alpha-tocopheryl acetate, and ROPUFA oil to the orange concentrate while stirring constantly. Dissolve the ascorbic acid and sodium benzoate in water, then mix both the solutions, one after the other, into the orange concentrate. Add the orange flavor. Homogenize the base at 250 bar, finally adding 1.00 g lime flavor and 0.42 g acesulfame.

Justifications for Dietary Fat Modification

Modification of ingredient characteristics and amounts from those traditionally and conventionally taught in the art can be justified on the basis of economics (lowered cost), organoleptic enhancements (better appearance, taste, etc.), improved functionality (easier processing, longer shelf life, etc.), and physiological characteristics (longer life or better quality of life for the consumer). No doubt there are other reasons that could be compelling in special circumstances. The emphasis in recent years on changing fat types and contents in the American diet has been almost exclusively directed toward the supposed physiological improvements the consumer can expect from the modifications. Based on limited experimental evidence and sometimes contradictory epidemiological data, nutritionists have been able to develop recommendations such as those shown in Table 7.6.

Table 7.6
RECOMMENDED DAILY CONSUMPTION OF DIETARY LIPIDS[1]

Type of fat or fatty acid	Typical current consumption		Recommended consumption	
	%	grams	%	grams
Total fat	40	115	24-28	72
Saturated fatty acids	15	43	<10	<30
Monounsaturated fatty acids	18	52	12-14	36
Polyunsaturated fatty acids	7	20	6-7	18
N-6 PUFA	6.5	18	5	14.5
N-3 PUFA	0.75	2	1.25	3.6
Linoleic acid	6.2	17.5	5	14
α-linolenic acid	0.7	2	1	3
EPA	0.03	0.1	--	--
DHA	0.03	0.1	--	--
DHA+EPA	--	--	0.27	0.8

[1] Notes:PUFA = polyunsaturated fatty acids; EPA = eicosapentaenoic acid; DHA = docosahexaenoic acid.
*Source:*From "ROPUFA: Food Enrichment" of Roche, based on NATO ASI Series A Life Sciences vol. 171, edited by C. Galli and A. Simopoulos (1991).

STEROLS

The term, "sterol," refers to a class of higher alcohols widely distributed in nature. When isolated, they are, in general, colorless crystalline compounds. They are nonsaponifiable and are soluble in certain organic solvents.

Plant Sterols

Sterols are not incidental substances, storage compounds, or waste materials. They are essential for the viability of eukaryotic cells. They are found in the free form or as esters conjugated to fatty acids. The concentration of free sterol determines the fluidity of eukaryotic cell membranes, but their esterified forms cannot participate in membrane assembly.

Several kinds of sterols are found in plants and are carried through to food ingredients and ultimately form part of the human diet. To name a few, campesterol, stigmasterol, and β-sitosterol. Sitosterol, of which there are several varieties, is found in relatively large amounts in wheat embryos, corn oil, and many other plant preparations.

Cholesterol is found in many animal tissues; its distribution and the amounts in which it is found in ingredients derived from animals will be discussed in considerable detail in a later section.

Table 7.7
TOTAL STEROL CONTENTS OF CERTAIN FATS AND OILS

Fat or oil	Sterols %	Fat or oil	Sterols %
Beef tallow	0.08-0.14	Olive oil	0.23-0.31
Butterfat	0.24-0.50	Palm oil	0.03
Cacao butter	0.17-0.20	Palm kernel oil	0.06-0.12
Coconut oil	0.06-0.08	Peanut oil	0.19-0.25
Corn oil	0.58-1.00	Poppyseed oil	0.25
Cottonseed oil	0.26-0.31	Rapeseed oil	0.35-0.50
Lard	0.11-0.12	Rice bran oil	0.75
Linseed oil	0.37-0.42	Sesame oil	0.43-0.55
Mowrah fat	0.04	Soybean oil	0.15-0.38
Mutton tallow	.03-0.10	Wheat germ oil	1.3-1.7

There seems to be an interrelationship of sitosterol and cholesterol absorption from the intestinal contents in man. An old paper by Beveridge et al. (1964) reports an experiment which showed a distinct lowering of cholesterol levels in human subjects dosed with sitosterol. The subjects consumed for seven days a homogenized formula providing 45% of calories as butter fat. The panelists were then divided into ten groups and continued on the butterfat regimen plus a commercial preparation of β-sitosterol in amounts of 50, 100, 200, 300, 400, 600, 800, 1,600, 3,200, or 6,400 mg per 950 kcal. Starting at the 300 mg level, the sitosterol caused progressively larger decreases in plasma cholesterol concentrations. The differences were statistically significant in all instances. It was concluded that the known hypocholesterolemic properties of corn oil could be attributed to the large amounts of plant sterols in this oil.

Cholesterol

Cholesterol is a steroid alcohol that is an essential component of the cell membranes in the bodies of animals, and it has other functions as well. Cholesterol is synthesized by animals; in fact, most tissues can make this substance, even though the most active sites of synthesis are the liver and the intestines. Therefore, a person who subsisted on a pure vegetarian diet (no animal products whatsoever), thus consuming zero cholesterol, would nonetheless have a substantial amount of cholesterol introduced into his bloodstream from endogenous sources.

Cholesterol is present in more than trace amounts. There is an average content of about 0.2% cholesterol in the human body, most of it in the skin, nerves (including brain), and muscle. It is also used as a raw material from which the body makes bile acids and certain hormones. From these, and other facts, it is obvious that cholesterol must be present for life to continue in the human body. One might be tempted to conclude that lowering cholesterol content of the blood to the lowest possible level would not necessarily be an unmixed blessing.

The esterification of intracellular steroids is mediated in mammals by a membrane-bound enzyme that seems to be an essential unit in the control of the concentration of free cholesterol in the blood serum. Increases in the activity of this enzyme change several pathways that can lead to hyperlipidemia and atherosclerosis. Sterol esterification modifies the action of the low-density lipoprotein receptor (LDL) and causes serum lipoprotein composition to be conducive to atherogenesis, and may be a rate-limiting step in absorption of sterol from the intestines. Deposition of cholesterol esters in the arterial wall is an important step in scarring of the vessel (from Yang, et al. 1996).

Cholesterol is transported in the blood from the liver to body tissues associated with certain lipoproteins, along with triglyceride fats. It is carried away from body tissues to the liver, there being mixed with bile acids and excreted into the intestines where they perform useful functions in the digestive process, including dispersing fats. Part of the cholesterol/-bile acid mixture is re-absorbed by the intestinal wall and recirculated. These features seem to indicate the existence of a homeostatic function that tries to regulate the blood concentration of cholesterol within a range that is presumably optimal for the organism.

The source and transport of cholesterol would be of only academic interest for the purposes of this book if it were not for the apparent connection of cholesterol with numerous circulatory disorders, and, in particular, atherosclerosis, a type of arteriosclerosis in which fibrous thickening of the arterial intima is connected with atheromatous degeneration.

Cholesterol and disease — Cholesterol has been implicated in the development of human atherosclerotic heart disease and other disorders of the circulatory system. It is fairly simple and inexpensive to determine the amount of cholesterol in the blood, the so-called serum cholesterol, and it is this datum that has attracted the most attention from scientists who have investigated the relationship of cholesterol to diseases of the heart and arteries.

It is easy to prove that the amount of cholesterol in the blood varies with the consumption of dietary cholesterol. It appears that there is a statistically significant relationship between the presence of high levels of serum cholesterol over long periods of time and the incidence of heart attacks, arteriosclerosis, and some other diseases. There is also no question that cholesterol is one of the major components of the obstructive plaques that are found in sclerosed blood vessels. These, and similar observations have led to recommendations of government agencies, public service organizations, and nearly all doctors that persons should seek to reduce their consumption of foods that contain significant amounts of cholesterol.

In the blood, cholesterol is transported in the form of large lipoprotein complexes. The lipoproteins may be classified on the basis of their electrophoretic mobility into alpha- and beta-lipoproteins. These two classes can be further subdivided on the basis of their density as determined by centrifugation; i.e., into high density lipoproteins (HDL) and low density lipoproteins (LDL). There is also a fraction called very low density lipoprotein (VLDL) which has not received much attention although it is claimed it also plays a role in arteriosclerosis and other circulatory disorders. These classifications have assumed some importance in medical publications, since it appears that the damage to blood vessels often attributed to "cholesterol" is associated with only certain of these combined forms of cholesterol, the other types being innocuous or even beneficial.

It is the current belief that the risk of incurring coronary heart disease, as well as other some other disorders of the circulatory system, tends to be related in a direct but not linear fashion to the content in the blood of low-density lipoprotein (LDL) and in an inverse but not linear pattern to the content of high-density lipoprotein (HDL) in the blood serum. High density lipoprotein appears to assist in removing cholesterol from cells and transferring it to the liver for excretion. LDL is involved in the transport of cholesterol into the cells and is believed to be harmful for this reason.

For people without detectable coronary heart disease, a total blood cholesterol level of less than 200 mg/dL is considered desirable; a level of 200 to 239 mg/dL is considered borderline, and 240 mg/dL is regarded as too high. An HDL level of less than 35 mg/dL is defined as "low" and is said to indicate an increased susceptibility to coronary heart disease. HDL readings of 60 mg/dL or more indicate a negative risk factor, that is, it protects against coronary heart disease.

The government and university savants who know what is good for us tell us that by the year 2000 all adult Americans should have reduced their cholesterol level to 200 mg/dL or less. It appears that about 12.7 million Americans will require drug therapy to achieve this level (Larkin 1994). The remainder can probably be safe merely by modifying their diets considerably. For those who prove to be refractory to dietary modification, lifetime commitment to drug therapy is the answer.

Bile acid sequestrants are approved for use by patients with high LDL-test results. Side effects are generally minor. Niacin products are approved for use in addition to dietary therapy for patients with high cholesterol levels or high triglyceride levels who do not respond adequately to diet and weight loss regimens. Side effects are not always minor; patients may incur irreversible liver damage, but never mind, they will still have low LDL.

Other treatments are being developed. Beitz (1995) patented a method whereby an enzyme is placed in food to convert cholesterol to coprostanol. The latter is a reduced form of cholesterol that is said to occur naturally in the large intestine. Unlike cholesterol, it is not taken into the bloodstream The enzyme favored by Beitz, cholesterol reductase, can be extracted from bacteria or even from green plants. The cited patent describes how it can be isolated, purified, then encapsulated into pill form. Another method might be to include bacteria that produce cholesterol reductase into foods, such as yoghurt, that normally include live bacteria.

Significance of serum cholesterol measurements — Of what value are measurements of serum cholesterol? Are they accurate enough to justify all of, or any of, the conclusions that have been reached about the results of diet and dietary modification? Irwig et al. (1991) remind us that an individual's blood cholesterol measurement may differ from the true level because of short-term biological variabilities and technical measurement uncertainties. Using data on the within-individual and population variance of serum cholesterol, they tried to find the answer to the question, "Given a cholesterol measurement, what is the individual's likely true level?" They found that the confidence interval for the true level is wide and asymetrical around extreme measurements because of regression to the mean. They say that, in general, *confidence levels are too wide to allow decision making and patient feedback about an individual's* cholesterol response to a dietary intervention, even with multiple *measurements*. And, "If no change is observed in an individual's cholesterol value based on three measurements before and three after dietary intervention, the 80% confidence interval ranges from a true increase of 4% and a true decrease of 9%."

As the reader will no doubt recognize, an 80% confidence level is not a figure on which one would ordinarily want to base serious behavioral modifying actions, and range would be even wider for confidence intervals of say, 90% or 95%. In other words, if the statistical analysis of Irwig et al. is correct, many clinical decisions are being made on the basis of essentially meaningless test results. The medical establishment, or some part of it, doesn't think this lack of significance is of particular interest.

Gwynne (1991) tells us that "Ambiguity does not arise when values are well above or well below the cutpoint." A certain amount of ambiguity does arise regardless of how far the value is separated from any point, there is no absolute certainty in any measurement. There is, of course, a greater likelihood that a test value far above a mid-point will be indicative of a true situation of being somewhere above the midpoint than a value only slightly separated from the midpoint.

Gwynne continues, "There is considerable variability in response to dietary modification. Some patients have such dramatic declines in cholesterol levels that posttreatment measurements clearly [sic] reflect the change. The average response is more modest, approximately 3% to 11% in free-living populations. Thus, many patients who have carefully followed a cholesterol-lowering diet may not exhibit a significant change in measured cholesterol level. The patient who claims to adhere to dietary advice but shows no change in measured LDL cholesterol level may well be telling the truth." Of course, and the doctor who relies on minor differences in test results as justification for changing the dietary habits of a patient may be ignoring more important indications to the patient's real problem which will remain untreated.

Epidemological studies are not entirely consistent with the theory that serum LDL levels and cholesterol consumption are the most important causes of coronary artery disease. Of those groups studied by Yoon (1996), Asian Indians have one of the highest rates of coronary heart disease in spite of their largely vegetarian diet, which limits cholesterol and saturated fat intake. And, risk factors such as high serum cholesterol levels, cigarette smoking, and hypertension are not disproportionately present in Asian

Indians. However, lipoprotein α-1 is elevated disproportionately, perhaps due to genetic influences.

Countless pieces of anecdotal evidence tend to show the lack of consistency in the theory that high cholesterol or high fatty acid intake is the cause of early development of life-threatening conditions in the circulatory system. Who. does not know an elderly, well-preserved person who has for several decades insisted on one or more eggs for breakfast, often accompanied by ham or sausage with gravy on the side? And, who takes their exercise in a rocking chair. Genetic factors are possibly overriding.

It is certainly known that only about half the heart-attack victims in the US are considered to have had high cholesterol and many of them did not smoke or have other risk factors commonly thought to be effective in causing heart attacks. We are justified in thinking that a very large piece of the puzzle is missing, and that, indeed, none of the puzzle has been put together properly. Rather typical of the popular media's reports along these lines is the discussion of a University of Georgia study of centenarians: "... most of them eat things like bacon, sausage, eggs even..." Some smoked for part of their lives. Not all had parents who were long-lived, or who even had average life spans (Cooper 1992). What is the answer? Whatever it is, we probably won't find it by performing more serum cholesterol tests.

Table 7.8
HIGHER SERUM CHOLESTEROL LOWERS CANCER RISK FOR MEN[1]

Serum cholesterol level	Incidence of colon cancer	Incidence of other cancers
mg/100 ml	%	%
126-189	4.6	13.4
190-219	1.7	10.5
220-249	1.3	9.8
250-279	1.5	9.2
280-545	0.4	7.2

[1] Subects followed over an 18-year period. Data are age-adjusted for men aged 35 to 69 years. The populations at risk vary from 242 for the lowest serum cholesterol level to 570 for the medium serum cholesterol level.

Source Adapted from Rawls (1981).

A further consideration, seldom brought into the picture when there are discussions of the supposed effect of serum cholesterol levels on heart disease, is the result of dietary changes on other diseases — the possible increases in other disorders that might result from lowering the amounts of cholesterol in the blood. The National Research Council's Food and Nutrition Board, in its recommendations for a wholesome diet for Americans (issued in 1980), said there is "no basis for recommending that healthy adults decrease dietary cholesterol."

High levels of serum cholesterol seem to reduce the risk of contracting colon cancer. In the famous Framingham study, which began in 1948 and continued for decades, it was found that colon cancer in men with the lowest levels of serum cholesterol (less than 190 mg per 100 ml) was about three times higher than for men with higher levels. The incidence of cancers other than colon cancer also decreased with increasing cholesterol level. There were no clear-cut effects in the women (Rawls 1981). Table 7.8 summarizes the relationship between serum cholesterol levels and cancer risk.

Are arterial plaques caused by blood cholesterol? — As this is being written, another piece of evidence appears that suggests coronary artery disease is caused by a bacterial infection (Winslow 1996, reporting work done at LDS Hospital and University of Utah School of Medicine). The bacterium in question, *Chlamydia pneumoniae*, was found in 73% of 90 patients who underwent a procedure to clear an obstructed coronary artery; when the investigators looked at similar arterial tissue obtained from 24 patients who did not have coronary artery disease, they found the bacterium in only one. Earlier, a cardiologist, Dr. Stephen Epstein, had suggested that a virus, cytomegalovirus, might infect and damage coronary artery tissue.

This sequence of events is reminiscent of the early stages of the discovery that bacteria were the cause of most peptic ulcers, a simple straightforward proposition that should have been investigated decades ago before tens of billions of dollars were spent, before hundreds of thousands of lives were lost, before millions of surgical operations were performed, trying to treat stomach ulcers with dietary regimens and with medications that affected only a symptom of the disease, if that.

Cholesterol reduction by dietary modification — Most authorities seem to agree that cholesterol is not found in any plant product, including vegetable oils. Some published analyses do, in fact, show traces of cholesterol or cholesterol-like substances in vegetable lipids but it is not believed this constitutes a practical or even a legal problem. Actually, a pure animal fat ("pure" in the sense that it contains only triglycerides) wouldn't contain cholesterol either, but lard, butter, and other common shortenings made from animal fats do include a small amount of this substance.

Results of a 21-day study with cholesterol-fed hamsters showed that rice bran and oat bran were effective in lowering plasma cholesterol, but neither wheat bran nor a blend of wheat and oat brans was effective. In a follow-up test, it was found that removing the lipid with hexane greatly diminished the ability of rice bran to lower plasma cholesterol. These data suggest that the lipid fraction of rice bran contains a significant portion of its hypocholesterolemic properties. Investigators at Louisiana State U. found in a crossover design experiment with mildly hypercholesterolemic (227 mg/dl) human subjects consuming 100 g of rice bran a day that diets containing 37% cal as fat and 300 mg cholesterol caused a 7% drop in total cholesterol and a 10% drop in LDL-cholesterol. The results are similar to those obtained using oat bran. Another group, using a non-crossover design with mildy hypercholesterolemic subjects consuming a rice bran/germ Product, were able to show decreases of 8% and 14% total and LDL-cholesterol, respectively. The persons conuming the oat product showed decreases of 13% and 17% for these values.

A group of Australian nutritionists compared the effects on mildly hypercholesterolemic men of consuming an additional 12 g of dietary fiber a day from wheat, rice, and oat brans. Both rice bran and oat bran resulted in a significant increase in the ratios of HDL-C:Total C and Apolyprotein-A:Apolyprotein-B. Both wheat and rice bran diets caused increased stool frequency, as compared to the effect of oat bran. Willett (1990) said, "The studies we have examined suggest that oat bran and rice bran lower serum cholesterol levels in hypercholesterolemic animals and humans under appropriate conditions. Data suggest that the bran in its entirety (fiber, lipid, protein, micronutrients, polyphenols, etc.) has a greater effect than any fraction. Rice bran data also suggest that the lipid fraction is involved with some small hypercholesterolemic properties remaining in defatted bran" (Betschart 1991).

Nicolosi (1991) found that rice bran oil has the ability to lower cholesterol levels up to 30% without reducing the so-called good cholesterol that is said to confer some protection agains heart attacks. He said, "Rice bran oil may produce these dramatic drops in harmful cholesterol by mechanisms that are not associated with its fatty acid composition like other vegetable oils." Rice bran oil is rich in unsaponifiable substances that appear to be responsible for its health effects. Two groups of monkeys were compared: (1) One group ate a "typical American diet" containing more fat than nutritionally recommended, (2) The second group ate a "healthy" diet containing more fat than nutritionists recommend. Depending on the amount of rice bran oil the monkeys were given, Nicolosi found decreases of 20% to 30% in LDL cholesterol levels. In some cases, he found a slight increase in HDL. LDL (low-density lipoprotein) has been implicated in hardening of the arteries and resulting heart disease, while HDL (high-density lipoprotein) removes cholesterol from the blood and possibly reduces the risk of heart disease.

Nicolosi claims that rice bran oil contains substances that help block the deposit of cholesterol inside arteries. The beneficial effects of rice bran oil are attributed to the effect of unsaponifiable oils in reducing cholesterol absorption from the intestines, in reducing the production of cholesterol in the liver, and in retarding oxidation that accelerates the deposit of cholesterol in the arteries.

It has been claimed that introduction of sucrose polyesters into the diet causes a reduction in serum cholesterol levels. Articles by Mattson et al. (1976) and Fallat et al. (1976) and the patent of Seligson et al. (1988) can be considered representative. Seligson and co-inventors disclosed a fat- and protein-containing food composition that provided at least one gram per serving of a non-digestible, non-absorbable sucrose fatty acid ester having at least four fatty acid ester groups, each fatty acid having from 8 to 22 carbon atoms. The food should also have a ratio of vegetable protein to sucrose fatty acid ester of at least 1.25 to 1. The possible role of Olestra (another sucrose polyester), elsewhere mentioned as a non-caloric frying medium, in lowering the serum cholesterol levels of persons consuming snacks cooked in the material is worthy of thought in connection with these patents.

One of the examples provided in the disclosure of Seligson et al. is a modified peanut butter composition containing 3.2 g sucrose polyester and 9 g vegetable protein in a serving of 32 g (2 tablespoonfuls). It is formulated as follows:

69.93% finely ground roasted peanuts containing 52% oil.
11.46% finely ground roasted peanuts containing 17% oil.
10.00% sucrose polyester.
 0.70% emulsifier
 5.8% sugar 12X
 1.20% salt
 0.40% peanut oil
 0.50% molasses

Baked products with reduced/no cholesterol — It would seem to be to the advantage of marketers of bakery foods that the naturally low cholesterol of most bakery products be emphasized in advertising, labeling, and other publicly disseminated information. Any foodstuff consisting entirely of plant-derived materials will be free of cholesterol. The cereal technologist is thus conveniently situated for designing low-cholesterol or no-cholesterol foods.

Formulating plain bread and rolls that are cholesterol free should not present major problems. This simply requires eliminating any animal fats or ingredients made from animal fats when formulating the product. None of these ingredients are essential to ordinary white bread or bread rolls. Thus, the formulation changes required to make cholesterol-free loaf bread and bread rolls are minimal, and any baker can easily make them with little or no effect on the finished product and no need for alterations in the production equipment provided the changes involve merely substituting vegetable shortenings having physical characteristics to those of the animal-fat shortening being replaced.

The producer must, however, exercise caution in screening minor ingredients and processing aids if a "no cholesterol" claim is being made. Some emulsifiers are made from animal fats and thus may contain traces of cholesterol. Pan greases, even those made primarily from mineral oil, may contain animal fats. Washes used to impart glossiness to the crust may include substances having some cholesterol.

Removing animal shortenings, including butter, and substituting shortenings derived from vegetable fats and oils for them, may not be a major formulation problem, but replacing egg and milk products in products where they contribute to flavor, texture, and label claims will be much more difficult. In fact, it is important to recognize at the outset that label claims of "milk bread" and "egg bread," for example, cannot be made without the inclusion of some sources of cholesterol in the products.

Problems and opportunities with eggs — Eggs are nutritionally valuable for several reasons. They are relatively inexpensive sources of high quality protein. Egg yolk phosphatidylcholine and sphingomyelin are sources of choline, which has important functions in brain development, liver function, and cancer prevention (see, for example Woodbury and Woodbury 1993). The sn-2 locus in yolk phospholipids is the primary site where long-chain omega-3 fatty acids are incorporated when chickens are fed fish oil or flaxseed, making such eggs promising additives for infant formula and other foods (Bringe and Cheng 1995).

Egg yolk contains about 31% fat and 1.3% cholesterol. This makes the consumption of ordinary eggs undesirable for consumers who are concerned about their cholesterol intake. Considerable success has been achieved in formulating liquid mixtures for scrambled egg preparation, and they can be found in the refrigerator cases of nearly all supermarkets. Most of these appear to be based on egg white combined with a number of other ingredients intended to furnish the functional, coloring, and flavoring properties of egg yolk. These non-egg ingredients generally include some sort of modified vegetable oil, an emulsifier, and color. Often a non-egg protein source, and sometimes flavor, are added.

The better examples of these products successfully replicate most of the properties of normal scrambled eggs (or omelettes) when they are properly cooked. Generally, they are somewhat blander than scrambled whole eggs, which may be either a positive or negative characteristic, depending on the tastes of individual consumers. Similar items are being offered to bakers, confectioners, and other food manufacturers for use as ingredients to replace whole eggs or egg yolks.

For some years, there have been projects in progress to reduce or eliminate entirely the cholesterol content of eggs so as to be able to provide ingredients that can be used in low-cholesterol or no-cholesterol foods. Various proposals have been made for modifying the feed given to laying hens so that the cholesterol content of their eggs will be reduced. In general, these have not been very successful. Havens (1992) patented a feeding regimen that includes adding a large proportion of dehydrated cabbage to the laying mash. The inventor states that the cholesterol content of eggs were reduced to about 341 mg/100 g as compared to about 500 mg/100 g in ordinary eggs.

Other studies have had as their purpose the reduction of cholesterol in fluid whole egg and egg yolks removed from the shell. Many patents have been granted on reduced-cholesterol egg products (such as frozen egg yolks) suitable for replacing ingredients made from normal eggs.

The patent of Athnasios and Templeman (1992) includes in its "Background" section one of the most complete reviews of prior art in sterol removal from foodstuffs that your author has seen. The patent in question describes a procedure by which cholesterol (and other sterols) can be removed from fluid mixtures by contacting the fluid with charcoal that has been activated by a specific set of conditions. More than 90% of the sterol compounds can be removed, and some of the saturated fatty acids are also removed. Unsaturated fatty acids are not adsorbed. The process is also said to be useful for removing cholesterol from butter oil.

A patent granted to Conte et al. (1992) discloses a cholesterol-reducing system suitable for eggs. An edible oil containing 5% or more monoglycerides is mixed with an aqueous solution of egg, and the mixture is heated to a temperature below about 70°C. High-shear stirring is applied until all the cholesterol has had an opportunity to contact the oil, then the aqueous phase is separated for use as an egg ingredient. From examples described in the patent, it appears that a cholesterol reduction of as much as 90% can be obtained.

A commercial product, tradenamed Eggcellent, contains 74% less fat and 90% less cholesterol than commercial egg yolk when reconstituted on an equal protein basis. It is made by extracting dried egg yolks with supercritical carbon dioxide, which dissolves neutral lipids and cholesterol without harming or removing many of the functionally and nutritionally important components. The product has been offered for use in reduced cholesterol scrambled eggs, baked goods such as pound cake, egg noodles, ice cream, custard, mayonnaise, and sauces. It has been claimed that 80% to

100% cholesterol reduction can be achieved in food products where similar ingredients have been substituted (on an equal protein basis) for regular egg yolks. Possibly, such products could qualify as "no cholesterol" (<2 mg per serving) or "low cholesterol" (<20mg per serving) if the saturated fat contents are less than two grams per serving (Bringe and Cheng 1995).

In the patent of Meibach et al. (1996), a passive trapping technique is used to remove cholesterol from eggs. A vegetable oil, preferably an oil containing a large percentage of polyunsaturated fatty acid, is combined with egg white permeate and egg yolk using a high energy, high shear mixing device. Cholesterol concentrates in the oil, which is then separated from the yolk by centrifugation. The cholesterol-reduced egg yolks are reconstituted with vegetable oil and concentrated egg whites to produce a "whole egg" product in which the cholesterol level is relatively low and the ratio of polyunsaturated fat to saturated fat is increased.

Awad and Smith (1996) used betacyclodextrins to remove cholesterol from eggs. Other inventors have used the same substance, but none have apparently reached the 95% cholesterol-removal described in the patent. Most previous disclosures have claimed only about 80% removal.

A modified whey, fortified with calcium at about 5% level, has been recommended for use as an egg yolk replacer in mayonnaise-type dressings (LaBell 1996). According to a study conducted by the manufacturer, the ingredient suppressed acid sharpness in the dressing, stabilized the emulsion, improved viscosity, and gave good mouthfeel properties such as fullness and body. Fat reductions of as much as 75% were said to be achievable.

Modified dairy products — The 3.5% or so of fat that fluid milk normally contains has been the subject of considerable unfavorable publicity because of its supposed health-damaging properties. Attention has been drawn to the average 219 mg cholesterol contained in 100 g of butter, as is basis. There has been interest in developing a cholesterol-reduced or cholesterol-free butter. Three principles that have been investigated are: binding cholesterol, solubilizing cholesterol, and converting cholesterol to innocuous substances.

According to the references available to the writer, it seems that only steam distillation technology is being used at present to manufacture cholesterol-free butter. One patented process heats anhydrous butterfat in a distillation column under vacuum. Volatile fractions, including cholesterol and certain other compounds are stripped out by water vapors, and condensed. The fat, with 95% of the cholesterol removed, is cooled and packaged. Of course, this is butterfat, not butter, and the flavor and texturizing product of the finished material are bound to be substantially different than the natural product. There are labeling problems as well.

Extraction with supercritical carbon dioxide has been studied, as has adsorption of the cholesterol on various substrates.

Fat-free "cream" cheese is readily available to food manufacturers and it has been incorporated into a number of commercial products. One producer of this ingredient published a formula and process for making fat-free cheesecake; this delicacy had the traditional egg content of 12.6%, (perhaps somewhat on the high side) replaced by an egg substitute, which makes the formula eligible for inclusion in this section. The formula is reproduced in Table 7.9.

The ingredient described as fat-free cream cheese was said to be based on casein and emulsifying salts treated with high heat and shear, with a new culture system used to develop the expected taste. To compensate for the higher water content of the finished product, stabilizers and hydrocolloids were added. In the baking process it was found to be necessary to adjust baking times, add steam to the oven, and increase cooling times to minimize cracking.

Table 7.9
EGG-FREE CHEESECAKE[1]

Ingredient	Traditional	Egg-free
Cream cheese, regular	60.8	--
Cream cheese, fat-free	--	51.7
Ricotta cheese	4.9	--
Yoghurt, nonfat lemon	--	13.4
Sugar	21.0	19.1
Vanilla	0.5	0.7
Whole eggs, fresh	12.6	--
Lemon rind, grated	0.2	0.4
Cornstarch	--	1.7
Flour	--	1.4
Salt	--	0.1
Egg substitute	--	11.5

[1] Amounts are stated on a batch percent basis.
Source: Attributed to Raskas Foods Co. (Anon. 1994D)

chapter eight

INCREASING FIBER CONTENT

Fiber fortification of foods has been one of the greatest marketing success stories of the food industry in recent years. Not only has fiber content, in the widest sense, become a desideratum of the thoughtful consumer, but various combinations and classifications of fiber have been promoted as being superior to all others. The field is now so cluttered with variations on the major theme that it is becoming confusing to the technically oriented observer and is doubtlessly more so to the less specialized consumer. This is the type of situation that can lead to a strong desire to abandon the whole confusing, complicated, and ever-changing universe of healthy foods and return to the more accessible and satisfying diet of beer and pretzels.

In preparing this chapter, it is the intent of the author to review justifications for the use of fiber (with emphasis on the physiological aspects of fiber consumption), to identify sources of fiber found in nature and available for use as ingredients, and to present a wide variety of formulas that have been proposed for fiber-enriched foods.

FACTORS ON WHICH BELIEFS IN FIBER BENEFITS ARE BASED

The belief in a healthgiving function of the indigestible portions of foods, especially of vegetables and grains, has been implanted in Western culture for a long time. Until the last few decades, this substance was called "roughage" by analogy to the hay, bran, and silage used in large amounts as feed for ruminants and called by that name. The meat-and-potatoes contingent disparagingly called it "rabbit food." According to Ory (1991), current interest in dietary fiber stems from the publications of Denis Burkitt, who sometime in the 1960s became convinced that an African diet of coarse grains, cereals, yams, and beans was the primary reason for the low incidence of colon-rectal cancer there. It it strange that Burkitt did not relate the extremely high incidence of cancers of lymphoid tissue among these populations to this diet.

At any rate, this was the starting point for the intensive research on fiber in the human diet which is still going on, and which has carried the claims of fiber benefits far beyond the early ideas of Burkitt.

Early Reports on the Beneficial Aspects of Dietary Fiber

Publications by medical researchers and nutritionists such as Cleave et al. (1969), Trowell (1972), and Burkitt (1973) seem to indicate that increased incidences of gastrointestinal and cardiovascular diseases (and perhaps cancer) are correlated with decreased intake of dietary fiber. The etiology of this relationship was and is obscure, but it has been shown clearly that transit times of food residues in the colon, stool weight and concentration, and the binding of bile salts in the intestines are directly affected by the amount of dietary fiber consumed. These factors singly or in combination are thought to result in more generalized kinds of physiological benefits when dietary fiber intake is increased.

Fiber and Cancer

Some recent studies present a confusing picture of the fiber:cancer relationship. Frühbeck (1996) briefly reviewed some of these. He says, "it is well known that foods with high fiber content are rich in phytic acid, saponins, tannins, flavonoids, phytoestrogens, [inhibitors of protease as well as amylase] ... lignans, and lectins ... Most of these factors are associated with a cholesterol-lowering effet, and therefore, with heart disease risk reduction." The so-called Scandinavian paradox, is, however, a fly in the ointment. This is a negative correlation between average fiber intake and age-adjusted annual rates of colonic cancer. Comparison of two closely related Scandinavian populations showed that Danes had a much higher incidence of colorectal cancer than Finns, although both groups consumed approximately equal amounts of dietary fiber. Further analysis, Frühbeck states, shows that the phytate intake was 20% to 40% higher in the relatively low risk Finnish population. Another variable, not mentioned by this writer, is the very different racial background of the Finnish and Danish people, which might indicate a genetic reason for the difference in cancer incidence.

In response to Frühbeck, Rimm et al. (1996) state that a further examination of the data, particularly by factoring in the amounts of flavonoids and phytates did not show very high correlations with coronary heart disease.

Fiber and Cardiovascular Diseases

In a more recent report, Rimm et al. (1996) tabulated the occurrence of fatal and non-fatal myocardial infarction during six years after the subjects filled out a detailed questionnaire about their dietary practices, including particularly questions about their usual intake of fiber and specific food sources of fiber. The results suggested an inverse association between fiber intake and myocardial infarction. The authors concluded, "These results support current national dietary guidelines to increase dietary fiber intake and suggest that fiber, independent of fat intake, is an important dietary component for the prevention of coronary disease." Within the three groups of dietary fiber foods involved (vegetable, fruit, and fiber) cereal fiber was the most strongly associated with a reduced risk of total myocardial infarction.

The rationale presented for their beliefs by proponents of the beneficial nature of increased amounts of dietary fiber in the Western diet is rather convoluted (not to say circular), and is based largely on demographic data, but some experimental evidence supporting it has accumulated over the last decade or so. For example, it appears that increased consumption of whole cereal grains (which are relatively high in fiber compared to many other foods) has decreased cholesterol levels in test animals. Nonetheless, the broader claims for health benefits, longer life, and improved quality of life still rely predominantly on statistical analyses and demographic data (Anon. 1985).

Animal studies have also shown that high intake of dietary fiber, and especially soluble fiber, decreases low-density lipoprotein cholesterol but has little or no effect on high-density lipoprotein cholesterol. The cholesterol lowering effects of a high-fiber diet are most apparent in dyslipidemic subjects.

As is their wont, the popular press latched on to the preliminary reports of possible fiber benefits and inflated their significance to the general health status of the average person far out of proportion. This storm of publicity, no doubt encouraged by those suppliers who expected to benefit from changing dietary patterns, whetted the appetite of consumers for foods of high fiber content and created opportunities for the marketing of foods specially formulated to fill the demand.

DIFFERENT TYPES OF FIBER

Since the principal reason for considering dietary fiber when choosing which of several available foods to consume is the health advantages that this component is expected to provide, it is justifiable to take into consideration the physiological effects of fiber when defining it rather than relying on chemical analyses alone. This reasoning has led to the widely accepted use of the concept of Total Dietary Fiber (TDF).

Physiological Aspects

"Total dietary fiber" is a rather inexact term intended to replace the even more inexact terms roughage, bulk, bran, fiber, and unavailable carbohydrates. A physiological definition of TDF describes it as the endogenous components of plant materials in the diet that are resistant to digestion by enzymes in the monogastric stomach and small intestine. Many of these substances are carbohydrates consisting of chains of monosaccharide units, cellulose being the most common example. These units are linked by chemical bonds called glycosidic bonds. The orientation of these bonds in the three-dimensional structure of the molecule is described as either alpha or beta, and can have an effect on the properties of the substance.

Enzymes secreted by the salivary glands and in the small intestine hydrolyze starch and other polysaccharides, yielding monosaccharide units which then enter the blood by passing through the wall of the intestine. These enzymes are specific for alpha linkages, so they do not affect carbohydrates based on beta linkages, and the latter enter, more or less intact, into the large intestine where some of them may be fermented by the bacteria always present there. Fermentation by-products include the gases causing flatulence, certain short chain fatty acids, and other substances that have undergone only limited investigation. In addition, the bacteria themselves increase in quantity.

Not all alpha-linked carbohydrates are completely broken down by the alpha-amylase and beta-glycosidase in the small intestine. For example, part of the raw (ungelatinized) starch may survive because the relatively large granules in which starch occurs in nature are difficult for the enzymes to penetrate, at least during the time available for action. Also, carbohydrates may contain ester, ether, or uronic acid groups that block the alpha-linkage sites, protecting them from attack by the enzymes. The sugar alcohols such as sorbitol and mannitol, if ingested in relatively large quantities may pass through the small intestine partly intact and behave like dietary fiber in the large intestine.

Common plant compounds that can be regarded as potential total dietary fiber include cellulose, hemicellulose, pectin, gums, mucilages, and lignin. The water insoluble fraction of these materials remains essentially unreacted, from a chemical standpoint, so they pass through the upper digestive tract and are available increase the bulk of the feces. Because the increased volume of the intestinal contents restricts access of enzymes to otherwise susceptible compounds and accelerates intestinal transit time, even hydrolysis of starch may be slowed and glucose absorption may be reduced if a large amount of fiber is ingested.

Water-soluble but indigestible substances absorb relatively large amounts of water, forming viscous liquids or gels in the intestinal tract, thereby slowing the passage of food, delaying gastric emptying time, decreasing glucose absorption, and, perhaps, lowering serum cholesterol levels. There are also negative effects experienced by some people, including a constant feeling of fullness or discomfort in the abdomen and a tendency to leak bowel contents.

Table 8.1
CHEMICAL CLASSIFICATION OF DIETARY FIBER

Fiber component	Chemical Components	
	Main chain	Side chain
Polysaccharides		
Cellulose (1,4 β-linked)	Glucose	*None*
Hemicelluloses:		
Arabinoxylan	Xylose	Arabinose
Galactomannan	Mannose	Galactose
Glucomannan	Galactose	Glucuronic acid
Pectic substances	Galacturonic acid	Galactose, glucose Rhamnose Arabinose Xylose Fucose
Beta-glucans*	Glucose	
Mucilages	Galactose, mannose Glucose, mannose Arabinose, xylose Galacturonic acid, rhamnose	Galactose

Gums	Galactose
	Glucuronic acid,
	  mannose
	Galacturonic acid,
	  rhamnose
		Xylose
		
		Fucose
		
		Galactose

Algal polysaccharides	Mannose Xylose Guluronic acid, mannuronic acid Glucose	Galactose
Non-polysaccharides		
Lignin	Sinapyl alcohol Coniferyl alcohol p-coumaryl alcohol	

* Beta-glucans are 1,3 β- and 1-4 β-linked and are found in appreciable amounts only in oats and barley. Pentosans are found in oat hulls and in the surface layers of other grains.
Source: Schneeman (1989)

Types of Fiber

There is no doubt that the tremendous publicity that has been given over the last ten or twenty years to the supposed beneficial effects of high fiber diets has convinced a large segment of the U.S. population that it would be in their best interest to consumer more fiber. Consequently, the fiber content of a food can be a selling point that has to be taken into account when the formula is developed and the label designed. First of all, the question to be asked is, "What is fiber?" as that word is understood by the consumer, the press, and regulatory authorities. There is more than one answer (see, for example, Schneeman 1986 and Schneeman 1989). A definition of dietary fiber arrived at by a committee is, "Endogenous components of plant material in the diet which are resistant to digestion by enzymes produced by man. They are predominantly nonstarch polysaccharides and lignin and may include, in addition, associated substance" (Anon. 1985).

Another definition is, " . . . the portion of plants which is not broken down by chemical action in the digestive system" (Harland and Hecht 1985).

The components of fiber that can be chemically characterized include the substances insoluble in water (cellulose, hemicellulose, lignin, insoluble pectins, insoluble gums, and pentosans) and the soluble fibers (soluble gums, soluble pectins, and polysaccharides not digested by amylases), according to Aaron and Stauffer (1986).

A chemical classification of dietary fiber appeared in an IFT committee report assembled and edited by Schneeman (1989). Some of the material in that report has been used for Table 8.1. As can be seen, the only non-polysaccharide of any consequence in this scheme is lignin, which is not customarily subdivided into other categories, largely due to its extremely heterogenous and mixed nature from a chemical standpoint, and the uniform resistance of all the forms to digestive attack. The carbohydrate groups of major importance include cellulose (by far the most predominant in nature), hemicellulose (a group defined largely by its response to extraction procedures), pectic substances, beta-glucans (of restricted occurrence but very important from the standpoint of health claims), mucilages and gums (overlapping and poorly defined categories), and algal polysaccharides (except for their origin, rather similar to mucilages and gums).

Crude fiber has long been measured by first removing lipids from a sample of plant material by solvent extraction, then digesting the product in dilute acid, then in dilute alkali, and determing the dry weight of what remains; the ash content is determined separately and subtracted from the dry weight. This is a test that has been applied for decades as part of the proximate analysis of foods The figures obtained for fiber in most common foods by this method are quite low, since only part of the cellulose and lignin (and none of the other dietary fiber materials) survive the extreme conditions of acid-alkali digestion. Another method for estimating crude fiber is arithmetical and simply involves subtracting from 100% the sum of the figures obtained in the other steps of proximate analyses. Neither method is suitable for measuring total dietary fiber, which often is five to fifteen times more than the crude fiber in a given sample.

Table 8.2
ANALYTICAL METHODS FOR DETERMINING FIBER[1]

Insolubles Determined	Solubles Determined
Crude Fiber Determination	
Two-thirds of cellulose	None
Acid Detergent Method	
Nine-tenths of cellulose	None
Lignin	
Small part of hemicellulose	
Neutral Detergent Method	
Cellulose	None
Lignin	
Hemicellulose	
Part of insoluble pectins	
Total Dietary Fiber Determination	
Cellulose	Soluble gums including β-glucan
Lignin	Soluble pectins
Hemicellulose	Soluble hemicellulose
Insoluble pectins	Polysaccharides that are not
Insoluble gums	digested by the enzymes used
Pentosans	

[1] From various sources.

One of the first "improvements" on crude fiber tests was the acid detergent fiber method first published in the 1960s. Samples are boiled in a solution of a detergent (cetyl trimethylammonium bromide) and then in sulfuric acid. The residue is filtered, dried, and weighed. A larger fraction of the cellulose and lignin content is measured by this procedure, compared to the crude fiber test. Lignin content can be determined by hydrolyzing the residue with 72% sulfuric acid and weighing the remnant.

The Neutral Detergent Fiber (NDF) procedure is regarded as an improvement but not as the final answer. It relies partially on isolating fractions by extractions with neutral and acidic detergents. Modifications of this method involving a preliminary hydrolysis of starch with alpha-amylase reduced problems that frequently occurred when analyzing samples containing substantial percentages of this carbohydrate. Omission of the acid and alkali hydrolysis steps leads to higher recoveries for nearly all of the fiber types.

Procedures of NDF type recover a substantial proportion of the cell wall constituents and are the basis for much of the data found in the large compilations of food fiber contents that are available for reference. However, water-soluble compounds such as gums, mucilages, and pectic substances do not appear in the residue, and thus do not contribute to the NDF figure though they play a role in physiological response. Other names for this system, or modifications of it, are Neutral Detergent Residue (NDR) and Insoluble Dietary Fiber (IDF).

A further improvement, regarded as essentially state-of-the-art, and the basis of an AOAC official method, can be summarized as follows. Duplicate samples are gelatinized in the presence of alpha-amylase, then digested with protein and with amyloglucosidase. These steps remove protein and starch. The soluble dietary fiber fraction is then precipitated with 78% ethanol in water, and the total residue is filtered, washed with ethanol and acetone, dried, and weighed. The ash and protein content of the residue are determined by the usual methods, and are subtracted from the residue weight and the difference reported as total dietary fiber. The residue will contain nearly all of the cellulose, hemicellulose, lignin, pectin, gums, mucilages, and modified cellulose. There have been refinements of the method in which an additional filtration step following treatment of the decantate with 95% ethanol is used to precipitate soluble fiber for further manipulations.

The so-called Southgate method requires a separate analysis for each of the fiber components (soluble and insoluble), and is said to be the most informative of all methods but it is very time-consuming and labor-intensive. It would seem to allow an unusual amount of cumulative errors.

FIBER IN NATURE AND COMMERCE

Although we can't be completely sure that we know what fiber is, we know where to look for it. Fiber is ubiquitous in the vegetable kingdom, not so in the animal kingdom. Whether it is found in the mineral kingdom depends on your definition. Wool is certainly a fiber by traditional definitions as is asbestos, but it is generally accepted in nutritional science and technology that the total dietary fiber content of any food or food ingredient derived entirely from animal sources or mineral sources is zero.

Numerous compilations of the dietary fiber content of selected foods have been published. USDA's Handbook 8, in its various editions and numerous volumes has more information than any other published source and it is available in machine readable forms. There are problems with this, and every other database of fiber content analyses. First, many of the samples tested are insufficiently identified, so the applicability of the values to any real-life situation cannot be substantiated in many cases. Secondly, it must be suspected that many of the averages include non-comparable samples, providing a spoonful of results out of a hash of samples that is, again, not very useful in predicting the composition of a sample collected from the marketplace. Thirdly, even the methods of analyses, the equipment used in the various laboratories, and the competency of the analysts, must be highly variable, especially in giant collections of data such as the USDA database. It must be emphasized, that the averaging of results obtained from variable samplings, variable methods, and variable personnel, is not statistically valid, and calculation of standard deviations in such cases is meaningless.

For readers who are interested in the inter- and intra-laboratory repeatability and precision of analyses for soluble and insoluble dietetary fiber, the study of Mongeau and Brassard (1990) reporting on a collaborative measurement of SDF and IDF by ten laboratories is worth consulting. They employed a rapid gravimetric method to determine the fiber content of seven products: corn bran, wheat bran, white bread, turnip, canned beans, rice, and whole wheat bread. The results of this study are too voluminous to repeat here, but the summary states the RSD_r (repeatability coefficient of variation) and RSD_R (reproducibility coefficient of variation) data from 8-10 laboratories ranged from 1.5% to 12.2% except for one product.

Of course, the legal status of information on the dietary fiber content of foods and food ingredients is based on different considerations, and can be quite definite and precise, even though it may bear little relationship to scientific concerns. Thus, when label declarations are being considered, and a source of information is needed as to the fiber content of, say, wheat flour, then recourse to the federal databases is a reasonable option.

A good source of dietary fiber determinations on cereal ingredients and cereal products based on analyses conducted in a single laboratory on presumably representative commercial products is the paper of Cardozo and Eitenmiller (1988). These data are more helpful in some respects than is the information in USDA Handbook 8 or the USDA database, although a considerable number of the Cardozo and Eitenmiller analyses have been entered into the USDA database.

Table 8.3

TOTAL DIETARY FIBER IN BREADS, MUFFINS, AND CAKES[1]

Product	Moisture	Total dietary fiber
	%	%
Boston brown bread	51.3	4.7
Bread, "Hollywood" light	37.8	4.8
Breadstuffing, flavored	60.0	2.9
Coffee cake, crumb topping	29.8	3.3
Coffee cake, fruit	35.6	2.5
Cornbread mix	34.4	1.8
Crispbread, lite rye	7.2	15.9
Croissant, butter	25.1	2.8
Croissants, cheese	21.0	3.8
Croissant, fruit	45.6	2.5
Croutons, plain	6.2	5.2
Croutons, seasoned	4.2	5.0
Danish pastry, fruit	27.6	1.9
Donuts, yeast, glazed	20.3	2.2
Egg bagels	33.0	3.9
Egg bread	30.4	3.8
French toast, frozen	48.1	3.1
Fruitcake	27.0	3.7
Gingerbread mix	38.5	2.9
Matzo, egg-onion	8.7	5.0
Matzo, plain	7.3	3.5
Matzo, whole wheat	4.8	11.8
Melba toast, plain	5.9	6.8
Melba toast, rye	6.7	7.9
Melba toast, whole wheat	6.1	7.4
Muffins, bran	35.0	7.5
Muffins, Eng., whole wheat	45.7	6.7
Norwegian flatbread	4.9	16.5
Pancakes, buckwheat	49.9	2.3
Pita bread, whole wheat	30.0	7.9
Taco shells	7.0	8.1
Tortillas, soft flour	20.2	3.7

[1] Values are averages of three samples taken from several mixed commercial samples; wet weight basis.
Source Cardozo and Eitenmiller (1988).

The scientists who performed these tests used the following method (essentially the procedure of Prosky et al. 1984). Samples were extracted with fat solvents if the fat content was 5% or more, then hydrolyzed with a heat-stable amylase preparation. After starch digestion was complete, the materials were treated with protease and amyloglucosidase to remove protein and any residual starch. Soluble fiber was precipitated by adding four volumes of ethanol to the digest, which was then filtered to collect the precipitate. The latter was rinsed with ethanol and acetone, dried, and weighed. Duplicates were analyzed for protein and for ash to allow correction for residual protein and ash. Total dietary fiber was calculated by subtracting the weight of residual protein and ash from the weight of the precipitate.

Fiber Content of Processed Foods

In Table 8.3 are listed the total dietary fiber contents of many bakery products and a few other finished products available in grocery stores, determined according to the aforementioned method o Cardozo and Eitenmiller and reported in the referenced study.

Table 8.4
TOTAL DIETARY FIBER IN RICE INGREDIENTS AND PRODUCTS[1]

Product	Moisture	Total dietary fiber
	%	%
Brown rice	11.7	3.9
Flour, brown rice	12.9	4.6
Flour, white rice	11.9	2.4
Glutinous rice	10.1	2.8
Instant rice, brand A	9.0	2.9
Instant rice, brand B	8.6	2.3
Medium grain rice	11.7	1.7
Parboiled rice, dry	10.4	2.2
Short grain rice	11.1	2.8

[1] Values are averages of three samples taken from several mixed commercial samples; as is basis.
Source Cardozo and Eitenmiller (1988).

Fiber contents for the products sampled for the data given in the preceding table are reported on an as-is basis, which is helpful for the ready-to-eat foods, but not as informative for items such as cornbread mix. For ready-to-eat products, the fiber content extends from 1.9 for fruit-dressed Danish pastry (heavy on fat and sugar, and fairly high in water content) to Norwegian flatbread (little or no sugar or fat, and low in water content). The weights of product in a serving are important, and would be expected to be relatively high for Danish pastry (perhaps four ounces), and low for the crispbread (likely an ounce).

Consumer types of rice products and most common ingredients made from rice are not particularly good sources of dietary fiber. Typical examples are described in Table 8.4. Even flour from brown rice, probably the least refined ingredient that could be described in an ingredient statement as "rice," contains only about 4.6% total dietary fiber on an as-is basis.

Table 8.5 includes one laboratory's reports of the total dietary fiber contents of certain pasta products. Whole wheat spaghetti, which has some major functional and textural differences from durum semolina pasta, and vegetable-flavored macaroni products (such as spinach and artichoke varieties) contain a moderate amount of TDF, but more common varieties are not particularly noteworthy sources of dietary fiber.

Table 8.5
TOTAL DIETARY FIBER IN UNCOOKED PASTA PRODUCTS[1]

Product	Moisture	Total dietary fiber
	%	%
Spaghetti, whole wheat	8.8	10.7
Spaghetti, spinach	8.7	10.6
Pasta, artichoke	9.7	9.8
Noodles, spinach-egg	8.5	6.8
Noodles, udon	11.5	5.4
Macaroni, protein added	8.5	4.3
Pasta, multicolor	10.1	4.3
Noodles, somen	11.3	4.3
Noodles, egg	9.5	4.0
Noodles, Chinese rice	6.2	3.9

[1] Values are averages of three samples taken from several mixed commercial samples; as is basis.
Source Cardozo and Eitenmiller (1988).

This low fiber content of pasta is to be expected since the common varieties are essentially pure durum semolina; it also follows that plain spaghetti, macaroni, vermicelli, and the thousand and one other shapes into which durum semolina doughs can be formed all have the fiber content of their base material (modified for moisture content, of course). So far as pasta as it is consumed is concerned, the sauces and adjuncts (meatballs, cheese, etc.) applied to it have large effects on the nutrient profile of the finished dish, as it is ingested by the consumer, to the extent that the alimentary paste can be considered as having significant nutrient effects only as a source of carbohydrates.

Pies, being relatively rich in sweeteners and shortening, and with many of the varieties being quite high in moisture, would not be expected to fall in the upper ranks of the list of fiber sources. This expectation is reinforced by the data in Table 8.6, which shows the dietary fiber content of a number of pies, cheesecakes, and similar products. None of these items contains more than 3.5% of total dietary fiber, and several pies fall within the range of 1% to 2%, on an as is basis. Consideration should be given to the portion size, however, since most pies are consumed in larger weights per serving than are items such as bread and biscuits.

Table 8.6
TOTAL DIETARY FIBER IN PIES[1]

Product	Moisture	Total dietary fiber
	%	%
Apple pie	59.9	2.8
Boston cream pie	47.6	1.4
Cheesecake, no-bake	46.2	1.9
Cheesecake, plain	45.4	2.1
Cherry pie	48.3	0.7
Chocolate cream pie	47.2	2.0
Egg custard pie	24.2	1.6
Lemon meringue pie	35.4	1.2
Pecan pie	23.2	3.5
Pumpkin pie	58.2	2.7

[1] Values are averages of three samples taken from several mixed commercial samples; wet weight basis.
Source Cardozo and Eitenmiller (1988).

In Table 8.7 are given the fiber and moisture contents of several biscuits, both sweet varieties (cookies) and savory/salty versions (crackers). The best source of fiber in this list is the whole wheat cracker, which has a dietary fiber content of slightly over ten percent, representing the effects not only of the bran and other highly fibrous portions of the wheat grain used as an ingredient but also the low moisture, which concentrates the fiber that is present in the finished product.

Table 8.7
TOTAL DIETARY FIBER IN
COOKIES, CRACKERS, AND CONES[1]

Product	Moisture	Total dietary fiber
	%	%
Crackers		
Cracker sandwiches*	3.9	1.1
Graham crackers	5.7	3.6
Honey graham crackers	5.6	1.7
Whole wheat crackers**	1.9	10.1
Cookies		
Butter cookies	3.5	2.4
Chocolate chip cookies	19.6	3.2
Chocolate sandwich cookies	5.0	3.0
Fig bars	19.9	4.6
Fortune cookies	8.0	1.6
Peanut butter cookies	11.5	1.8
Shortbread cookies with nuts	5.5	1.8
Vanilla cream cookies	2.9	1.5
Ice cream cones		
Brown sugar cones	2.9	4.6
Wafer cones	5.6	4.1

[1] Values are averages of three samples taken from several mixed commercial samples; wet weight basis.
* Peanut butter and cheese filling.
** "Triscuit"(tm).
Source Cardozo and Eitenmiller (1988).

Cookies can be expected to be highly variable in fiber content, with items such as cream-filled sandwich cookies and sugar wafers containing very little fiber of any type and molasses raisin cookies containing a substantial amount. Furthermore, different samples of the same-named variety can vary widely depending on the formula, a Garibaldi biscuit would contain a relatively large amount of soluble dietary fiber (from raisins) while a cheap raisin cookie might have only slightly more fiber content than a plain sweet biscuit.

Two ice cream shells, or cones, are included in Table 8.7. They contain an intermediate amount of fiber, on a percentage basis. In terms of its effect on the total diet is concerned, it must be understood that the portion size of the shell is quite small (ordinarily less than an ounce) and is, of course supplemented before consumption with a much larger amount of ice cream, which would contain very little fiber, even if was a variety with nuts. One can hardly expect to increase one's dietary fiber content appreciably by consuming an extra amount of ice cream cones.

Table 8.8
TYPICAL FIBER CONTENTS OF WHOLE GRAINS[1]

Ingredient	Total dietary fiber	Crude fiber	Water
Amaranth	15.8	3.77	9.84
Buckwheat [groats]	NA	1.76	8.41
Barley	17.2	2.85	9.44
Corn	NA	2.90	10.37
Millet	7.5	1.03	8.67
Oat groats	11.6	6.90	8.22
Rice [long grain brown]	3.5	1.32	10.37
Rye	NA	1.50	10.95
Sorghum	NA	2.40	9.20
Triticale	18.2	2.60	10.51
Wheat, HRW	12.6	2.29	13.10
Wild rice, raw	5.2	1.44	7.76

[1] Energy amount per 100 grams of edible matter. Fiber amount stated as grams per 100 grams, as is.
Source: U.S. Agriculture Handbook 8-20 (1989), Vollendorf and Marlett (1991), Marlett (1991), and elsewhere.

Fiber Content of Processed Ingredients

Whole grains are among the first natural ingredients coming to mind when reviewing the materials available for use in supplementing foods.

These ingredients are generally inexpensive, have "natural" connotations that make them highly acceptable to many potential customers, and have been used for so long and by so many people that the risk of unexpected health problems arising is negligible (however, look elsewhere for an explanation of the relation of celiac disease to the consumption of wheat and allied grains). Also, they are acceptable by all, or virtually all, groups that obey dietetic restrictions based on philosophical or religious tenets. Functionally and organoleptically, they may not be quite so desirable, however.

Wild rice, which is not even of the same genus as ordinary rice, is included in the following table. Like its namesake, it is fairly low in total dietary fiber content. Amaranth, a grain but not a cereal, has a respectable fiber content as does barley, even the pearled form of barley. Millet is about intermediate between rice and barley, while triticale (a hybrid of rye and wheat) is relatively high in dietary fiber content, at least according to this reference. Even so, none of these whole grains, or the meal ground from them, would be of much value as a source of fiber enrichment, unless they could be used as the principal ingredient, e.g., in granola bars, breakfast cereals, and the like.

Flaxseed, not included in the above table, contains about 40.4% total dietary fiber on an oil-free (oil normally about 32% to 38% dwb) dry-weight basis. The natural toxicants, cyanogens, are found in flaxseeds, as in some other foods, such as lima beans; in the body they are converted first to hydrogen cyanide, then to thiocyanate. The latter compound inhibits iodine uptake by the thyroid gland. It has been reported that thiocyanate is destroyed by cooking conditions. For example, the experimental high-fiber muffin formula shown below resulted in product which contained no detectable cyanogenic glucosides (Cunnane et al. 1993). Further details can be found in the booklet by Vaisey-Genser (1994).

960 parts wheat flour
600 parts ground flaxseed
658 parts fluid whole milk
300 parts whole egg
300 parts honey
  96 parts baking powder
  30 parts corn oil
  17 parts salt

Processed grains as fiber sources — As stated previously, whole grains are generally fair sources of fiber, mostly of the insoluble type. But, when grains are processed and fractionated, as in wheat milling, those parts containing the highest content of fiber tend to end up in animal feed while materials intended for human consumption will have relatively low contents of dietary fiber. This generalization is particularly appropriate for flour milling, but is not as applicable to preparation of oats for hot breakfast cereal use. Culinary oats is for all practical purposes the whole groat in various physical forms (flaked, rolled, cut, etc.) with most if not all of the seedcoat, or bran, present. Table 8.9 includes dietary fiber information on milled products made from common grains, both cereals and non-cereals.

Table 8.9
TOTAL DIETARY FIBER OF MILLED GRAINS [1]

Product	Moisture	Total dietary fiber
	%	%
Wheat products		
Flour, all-purpose, brand A	11.2	5.6
Flour, all-purpose, brand B	11.3	5.3
Flour, all-purpose, brand C	12.9	2.3
Flour, cake	11.6	3.7
Flour, high gluten	14.3	4.1
Whole wheat flour	11.5	14.6
Semolina	11.7	3.9
Rye products		
Dark rye flour	11.1	32.0
Light rye flour	10.6	14.6
Medium rye flour	8.8	14.7
Corn products		
Corn flour	10.9	13.4
Cornmeal	11.0	3.6
Corn meal, degermed	10.3	9.5
Masa harina	11.3	9.6
Other, or Mixed		
Amaranth flour	13.0	10.2
Buckwheat flour	12.3	10.0
Oat flour	12.2	9.6
Roman Meal (tm)	9.1	18.0
Triticale flour	9.4	14.6

[1] Values are averages of three samples taken from several mixed commercial samples; as is basis.
Source Cardozo and Eitenmiller (1988).

When cereals and other seeds are processed into food ingredients, the products generally fall into high fiber fractions, which until recent years have been regarded generally as byproducts more suitable for animal feed than for human consumption, and the more refined, lower fiber content fractions that find their way into commerce as food ingredients. Thus, we have wheat being processed into millfeed (bran, shorts, etc.) and flour, and corn being separated (by wet-milling) into gluten feed and cornstarch. With the craze for fiber enhancement of foods, a change occurred, and there developed a large demand for food ingredients of high fiber content, making some of the byproducts as valuable as many of the refined materials. Of course, it was often necessary to further refine or otherwise process the byproducts to make them suitable for human consumption, since little effort was formerly devoted to producing sanitary and contaminant-free feed constituents, cattle not seeming to mind an occasional insect part in their rations. In addition, it was often found in experimental programs that addition of an unmodified byproduct to bread (for example) had unfortunate effects on the appearance, texture, flavor, and storage stability of the finished food.

As we can see in the accompanying Table 8.10, ingredients obtained by milling grains (cereal and non-cereal) exhibit a rather wide range of fiber contents, with the whole grain products having substantially more fiber than the refined products; it can also be observed (though not from the table) that the latter are typically whiter in color, finer in particle size, and milder in flavor and odor than the byproducts.

Table 8.10
TYPICAL FIBER CONTENTS OF PROCESSED GRAINS[1]

Ingredient	Total dietary fiber	Crude fiber	Water
Barley, pearled	15.6	0.74	10.09
Cornmeal, degermed	5.2	0.62	11.59
Cornstarch	0.9	<0.1	8.32
Farina	2.7	0.2	10.50
Oatmeal	10.3	1.1	8.80
Rice flour, brown	4.6	1.29	11.97
Rye flour, medium	14.6	NA	9.85
Triticale flour, whole	14.6	1.50	10.01
Wheat, all-purpose flour	2.7	0.26	11.92
Wheat, whole grain flour	12.6	2.10	10.27

[1] *Source*: U.S. Agriculture Handbook 8-20 (1989), and elsewhere.
Fiber expressed in grams per 100 grams, as is.

Some of the milling byproducts that are mostly used for feeding cattle have quite high fiber contents (see Table 8.11). In commercial parlance, many of these by-products are called "bran." Bran is not a well-defined material, and its composition can vary widely from mill to mill and, indeed, can be changed by varying machine settings in the same plant. Ruminants (cows, sheep, etc.) can process most of this fiber into energy sources, in cooperation with the symbiotic bacteria that function in their gut. Humans are less fortunate, and excrete a large part of the fibrous material substantially unchanged. This disadvantage has been turned into an advantage through the magic of public relations, and we now "recognize" or at least accept the dogma that more fiber is good for us.

Table 8.11
TOTAL DIETARY FIBER IN MILLING BYPRODUCTS[1]

Product	Moisture	Total dietary fiber
	%	%
Corn bran	6.2	82.4
Oat bran	7.8	17.9
Rice bran	8.2	21.7
Rice polish	12.0	15.3
Wheat bran	11.6	42.6
Wheat germ, crude	9.5	16.9
Wheat germ, toasted	5.8	16.1

[1] Values are averages of three samples taken from several mixed commercial samples; as is basis.
Source Cardozo and Eitenmiller (1988).

Bran and most other milling byproducts are relatively inert so far as mild processing steps are concerned, and, in fact, consist largely of materials that are insoluble at normal temperatures. They are usually contaminated with bacteria at levels several times higher than, say, flour.

In a more recent reference, Koch (1993) lists the following total dietary fiber contents for some seed derivatives, including a number of non-cereal raw materials. He also gives soluble (included in above amounts for total) as 1% for corn bran, 4% for pea fiber, 4% for soy cotyledon flour, 5% for oat bran, and 64% for guar gum; significant amounts of soluble dietary fiber are not found in the ingredient materials called soy hull fiber, oat fiber, and cellulose gum.

Oat bran	17%
Wheat bran	44%
Guar gum	78%
Soy cotyledon fiber	80%
Corn bran	88%
Pea fiber	90%
Soy hull fiber	93%
Oat fiber	96%
Wheat fiber	98%
Cellulose gum	99%

Koch also says, "A review of the proposed fiber benefits indicates that gastrointestinal response can play a major role in the delay of colon or rectal cancer, treatment of constipation, or diverticulosis. Cholesterol reduction impacts heart disease, while diabetes is improved by hypoglycemic rebound [obesity reduction can be aided by the reduction in total calories of products consumed by diabetics]."

Oat hull, which is seldom if ever used in foods, contains about 98% dietary fiber, virtually all of this being insoluble fiber. Ramaswamy (1991) patented a method for making from oat hulls an extremely white powder having a high percentage of total dietary fiber (91.4%) and a low concentration of lignin (0.6%), and silica (ash 1.46%); crude fiber is 69.2% and pentosans 12.5%. The process involves subjecting ground oat hulls to alkaline digestion at elevated temperatures and pressures, then filtering, neutralizing, and bleaching.

Oat bran, the separation of which requires a substantially different process than normally used in wheat milling, contains about 7.4% insoluble fiber and 4.9% soluble fiber to make a total of 12.3% total dietary fiber. Typical rolled oats and flour, which contain bran, will analyze as 9.6% total dietary fiber, of which 5.8% is insoluble and 3.8% soluble.

Fiber contributions from other ingredients — "Other ingredients," as the term is used in this title, means highly modified grain products as well as ingredients not derived from grains.

The current emphasis on increasing the amount of fiber in human diets has led to the publication of numerous studies on means for incorporating various sources of these substances in bakery foods. For example, when a wheat fiber preparation containing 98% total dietary fiber was blended with a bread flour, with or without vital wheat gluten and sodium stearoyl-2-lactylate, and the mixtures used to prepare yeast-leavened loaves, it was shown that absorption and mix time increased as the wheat flour amount decreased. High-fiber breads containing wheat fiber exhibited reduced loaf volumes, pale crust color, and a coarse, open and nonuniform grain. Vital wheat gluten and sodium stearoyl-2-lactylate had no effect on loaf volume, but the latter improved both crumb grain and softness (Park et al. 1995).

Table 8.12 includes a few cereal and non-cereal products that may have utility in some specialized products. Only bulgur, which should have about the same fiber content as the whole wheat from which is made, has enough fiber to be suitable as an enrichment.

Table 8.12
TOTAL DIETARY FIBER IN MISCELLANEOUS
PRODUCTS AND INGREDIENTS[1]

Product	Moisture	Total dietary fiber
	%	%
Arrowroot flour	11.4	3.4
Bulgur	10.4	18.3
Cornstarch	7.8	0.9
Couscous	8.9	9.8
Hominy, canned	83.0	2.5
Tapioca flour	11.0	0.9

[1] Values are averages of three samples taken from several mixed commercial samples; wet weight basis.
Source Cardozo and Eitenmiller (1988).

Since an increase in fiber content seldom has an improving effect on the appearance, flavor, or texture of a product, the usual justifications for adding ingredients to enhance fiber content are either to decrease ingredient costs or to improve the healthfulness of a product. On the other hand, thickening and gelling agents (mostly regarded as soluble fiber) are essential as texture-modifying constituents in many dietetic foods. For example, these materials are functionally important ingredients in fat-free or fat-reduced salad dressings.

Modified grain products — Rice bran, unmodified, is not a good choice for fiber supplementation because it has a tendency to become noticeably rancid in a relatively short period of time. This difficulty can be overcome by appropriate processing methods, and various "stabilized" rice brans are being made and offered to food processors. One method for obtaining increased shelf life, is described in the patent of Hammond (1994); briefly, the inventor stabilizes rice bran by applying an anti-lipase enzyme that inactivates the naturally occurring lipase enzyme in the bran, mixing the rice bran with water, and allowing sufficient time for a wet, stabilized rice bran to be produced without denaturing the product. It is believed that most stabilizing methods rely on the use of moist heat to inactivate the troublesome enzyme. Table 8.13 lists three different types of processed rice bran that might be considered by formulators as fiber sources for dietetic foods.

Table 8.13
STABILIZED RICE BRAN PRODUCTS
FOR FIBER SUPPLEMENTATION[1]

Characteristic	Stabilized natural rice bran	Low fat rice bran	Defatted rice bran
Protein, %	13.0-16.0	15.0-18.0	16.0-20.0
Oil, %	18.0-22.0	6.0-8.0	0.5-1.5
Moisture, %	5.0-8.0	5.0-8.0	5.0-8.0
Ash, %	6.0-9.0	7.0-10.0	7.0-11.0
Crude fiber, %	6.0-9.0	7.0-10.0	7.0-11.0
Dietary fiber, %	25.0-35.0	27.5-37.5	30.0-40.0
FFA, %	1.5-2.5	1.5-2.5	1.5-2.5
Color	Tan	Light tan	Creamy tan
Flavor	Sweet. sl. toasted.	Sweet, sl. toasted.	Bland
Total plate count	<5000/g	<5000/g	<5000/g
Granulation	Flakes	40-mesh	100-mesh

[1] These data were provided by California, Inc., and relate to the series of products identified by the tradename "Calbran."

Other supplements that might be considered are purified cellulose, which should be close to 100% fiber on a dry weight basis, pectin, lignin, and various gums such as guar.

The author's experience suggests that any of these supplements, when included in formulas in more than trace amounts, have the potential of impairing the flavor, texture, and appearance of the finished product. Highly flavored, dense cookies are one of the best mediums for delivering fiber in a form acceptable to consumers. From an operational standpoint, corn bran is perhaps the best choice for a fiber supplement in cookies of this sort, especially if cost is a significant consideration, but it, too, can cause substantial problems resulting from changes in the physical properties of intermediates and the flavor and texture of the finished products. A type of corn bran specially prepared for use as a fiber supplement in foods contains 80% to 85% total dietary fiber (17% crude fiber), 10% starch, 2% fat, 5% protein, 1% ash, and 10% moisture, and contributes about 0.5 calorie per gram.

Among the cereal fiber concentrates commercially available are (with percent dietary fiber in parentheses, according to Hegenbart 1992): corn bran (85%), oat bran (17%), rice bran (27%), soy cotyledon (80%), and wheat bran (44%). The soy cotyledon material, a recent introduction, is said to consist of fiber extracted from the seed itself (not the hull) of soybeans.

Gums and mucilages — The following ingredients and their uses are mentioned in the review of Sanderson (1996); that author calls all materials in the following list, "gums," but it seems better to reserve that term to describe a more restricted group of natural and synthetic products, at least in the context of the present discussion.

Most of the materials in the list would qualify as dietary fibers, but a few would not. Starch and gelatin, for example are almost entirely digested when used as ingredients in processed foods and consumed by persons having normally functioning digestive systems. Resistant starch has unusual qualities which will be discussed separately.

1.*Thickeners*

1.1. Starch and starch derivatives — "by far the most widely used hydrocolloid in the food industry."

1.2. Cellulose derivatives — the most important is carboxymethyl cellulose, a cellulose ether, a term also used for methyl- and hydroxypropylmethylcellulose. Produces extremely clear solutions. Hydrates and builds viscosity quickly.

1.3. Cellulose — microcrystalline cellulose, tradenamed Avicel. Originally designed and still used as a fat replacer. Usually requires other hydrocolloids to supplement it for best activity.

1.4. Guar gum — the most widely used thickener, primarily because of its low cost. A galactomannan derived from seeds. A polymeric backbone of mannose residues that is water-soluble as a result of single-unit galactose side chains. More cold water soluble than locust bean gum and like it, interacts with xanthan.

1.5. Konjac — a mannan gum, chemically a polymer of glucose and mannose, with acetate as sidechains. Forms gels when interacting with xanthan or kappa-carrageenan. When deacetylated, it self-associates to form the gels that are common in many Asian cuisines.

1.6. Tamarind seed gum — stable viscosity over a wide pH range and in the presence of salt. Flow properties more Newtonian than other gums.

1.7. Tragacanth, karaya, and arabic gums — exudate gums.

1.7.1. Karaya gum — produces very high solution viscosity and good adhesive properties. It was widely used for salad dressings prior to the availability of xanthan gum.

1.7.1. Gum arabic — the most important exudate gum for foods. Solutions of up to 50% can be prepared. At low concentration, provides very little viscosity. Emulsifying properties due to the presence of protein attached to the polysaccharide (gums do not generally function as emulsifiers).

1.8. Xanthan gum — produced by the microorganism *Xanthomonas campestris*. Became available in the 1960s. It is a modified cellulose chain, having three sugar unit side chains. Produces high transient viscosity at low concentrations, but the gels are readily broken down by shear. The high viscosity tends to persist at high salt or acid concentrations and at elevated temperatures.

2. *Gelling agents*

2.1. Gelatin — the most widely used gelling agent. Gels melt around 30°C, and firm up at room or refrigerator tempoerature. Used in many foods.

2.2. Pectin — a complex colloidal carbohydrate made up of molecules containing a large proportion of units derived from galacturonic acid. It is commercially extracted from citrus peel and apple pomace. Pectin yields viscous solutions with water and is responsible for the typical structure and texture of jams and jellies. In dried form, it is a white amorphous powder. There are two broad types available for use as ingredients: high methoxyl and low methoxyl pectin, separated at about 50% methoxyl content. High methoxyl pectin requires pHs below about 3.5 and high soluble solids (more than 60° Brix) to form gels. Low-methoxyl pectin forms gels by interchain association induced by calcium ions, a reaction that is more or less independent of the sugar concentration. When completely, or almost completely, deprived of methyl groups by controlled hydrolysis, pectinic acid and methyl alcohol are formed.

2.3. Carrageenan — from red seaweed. Polysaccharide mixture, containing among other species, kappa, iota, and lambda, differing in their respective ester sulfate and 3,6-anhydrogalactose contents. Kappa and iota forms are primarily gel-formers, while lambda is mainly a thickener. Gelation is usually caused by heating and cooling, and is mediated by cations (potassium for k-gels and calcium for i-gels). Carrageenans can stabilize casein in milk products

2.4. Agar — from red seaweed. More common in Far East than in Western cuisine. In Asia, it is considered a food or food ingredient.

2.5. Alginates — from brown seaweeds. Linear polymers composed of mannuronic and guluronic [sic] acid. Their most important food property is their ability to interact with calcium ions, producing useful increases in viscosity at low levels of calcium, and with adequate levels of Ca, produce gels. Slow release of calcium, needed to produce optimal gels, can be had by use of sequestrants, typically sodium citrate, to compete for the cation. Can also form precipitates with acid, undesirably in some cases. Propylene glycol alginates have good acid stability and are used in beverages, syrups, salad dressings, and beer.

2.6. Gellan gum — a bacterial polysaccharide. Fairly recent development. Has been used or suggested for use in icings, fillings, dessert gels, low-sugar jams and jellies, puddings, and confections. Gel clarity, outstanding flavor release, and acid stability are other useful features. Gel formation, usually brought about by cooling, is induced by cations including hydrogen ions from acid.

Partially hydrolyzed guar gum, one version having the trade name "Benefiber" is described as being water soluble (giving solutions of low viscosity), colorless, flavorless, and odorless. Also, it is said to be stable to heat, acid, salt, and digestive enzymes. The manufacturer claims it can be used in fat-free blueberry muffins and sugar cookies, therein acting as a humectant to produce a moist and tender texture (LaBell 1995).

Cellulose fiber — The most concentrated form of fiber is purified cellulose. Some versions of this material will analyze 99% insoluble dietary fiber. For many years, highly purified cellulose material obtained from wood has been available for use as a food ingredient. The commercial items have several advantages. They are virtually flavorless, have unlimited storage life, and, since they never become sticky (even when damp) they can be dispensed without major problems. However, their inert nature can interfere with the formation of the delicate structures involved in dough and batter formation and in processing of other foods.

Within the last few years, cellulose fiber from other sources has become available. Sugar beet fiber, sugar cane fiber, and the like are largely cellulose, though generally not quite so pure as some varieties of wood cellulose; this may not matter in many applications. The fuzz that is retained on cottonseed after the ginning process (called linters) is now being processed into a food additive. According to Forand (1991), an improved material of this sort allows the production of baked goods having higher volume than those made with wood cellulose. The following formula for reduced calorie white pan bread was given by Forand (amounts in grams, unless otherwise indicated).

Sponge

490 bread flour of 12.8% protein.
30 vital wheat gluten

0.8 mineral yeast food
25 compressed yeast
320 water

Dough

210 bread flour of 12.8% protein
 54 vital wheat gluten
182 cellulose powder
 49 granulated sugar
 19 salt
  1.4 cellulose gum (CMC-7HF)
  1.4 calcium propionate
  3.6 crumb softener (GMS-90)
 69 ppm ascorbic acid
556 water

The water retention capabilities of purified cellulose fiber vary according to source of the fiber and its processing history. Typically, cellulose can absorb 3 to 10 times its weight of water. This water-absorbing power may effectively increase the concentration of any hydrocolloids that are present in the aqueous phase, leading to increased viscosity of the total system. Nonetheless, cellulose fibers by themselves do not ordinarily have much thickening power in solution; they tend to settle out. It is claimed, however, that cellulose fibers in combination with xanthan gum, guar, and carboxymethylcellulose thicken solutions synergistically. A reduced-calorie pancake formula incorporating the wood cellulose derivative Solka-Floc (of which there are several types) has been published (Anon. 1993):

Part 1

Solka-Floc 200FCC, 6.5%
Solka-Floc 900FCC, 0.5%
All-purpose flour, 17.9%
Baking powder, 1.8%
Baking soda, 0.2%
Sugar, 3.6%
Salt, 0.6%
Bealite Emulsifier 3401-L, 0.2%

Part 2

Powdered buttermilk, 5.00%
Polydextrose, 2.00%
Xanthan gum (Keltrol) 0.06%
Whole eggs, 8.00%
Water, 51.64%
Vegetable oil (Crisco), 2.00%

A high fiber, reduced calorie, molasses cookie can be made as follows. Cream 4.5 parts each of shortening, emulsifier, and sugar. Add 4.5 parts whole egg, 21.8 parts molasses, and 19.0 parts water, and beat until well blended. Mix separately 7.5 parts Solka-Floc 300 FCC, 0.9 parts Solka-Floc 900 FCC, 22.5 parts all-purpose flour, 7.5 parts polydextrose (Pfizer), 0.4 parts salt, 0.2 parts each of ginger, cinnamon, and xanthan gum, 0.7 parts baking soda, 0.9 parts nonfat dry milk, and 0.1 parts each of citric acid and nutmeg. Slowly add dry blend to the wet mix, blending until a uniform mixture is obtained. Deposit on greased baking sheet or on parchment and bake at 375°F for 12 to 14 minutes (Anon. 1992C).

Purified fibers from fruits and vegetables — Residues left behind in the processing of fruit and vegetable juices have been extensively investigated for possible application as enhancements for the fiber content of foods. The composition of these concentrates depends not only on the source of the raw material but also on the method used in separating the fiber and processing steps applied to it afterwards. As a result of these investigations, we find that low moisture apple and pear fibers have become available.

Table 8.14
SPECIFICATIONS OF APPLE AND PEAR FIBERS

	Low moisture apple fibers	Low moisture pear fibers
Moisture, %	0.5	0.5
Protein, %	7.2	9.1
Fat, %	3.1	1.8
Ash, %	1.3	1.4
Crude fiber, %	42.5	40.0
Total Dietary Fiber	57.7	56.0
Insoluble D.F., %	56.0	54.5
Soluble D.F., %	1.7	1.5
Neutral Detergent Fiber, %	63.2	61.0
Pectin, %	4.5	4.5
Bulk density, g/cc:		
Loose	0.45	0.49
Packed	0.59	0.59
pH, 1% solution	4.3	4.4
Calories, per 100 grams*	177.	177.

*Calculated using zero calories for Total Dietary Fiber.
Source: Product Data Sheet, TreeTop (Anon. 1987A).

Materials described as apple and pear fibers for food supplementation consist primarily of mixed cell wall constituents such as cellulose, hemicellulose, lignin, and pectin. In commercial practice, these ingredients are prepared from the solid residue remaining after juice processing. The raw material for low moisture versions is said to be relatively free of extraneous peel, stems, and seeds and other core materials, and is usually rather bland.

Oat bran and oat hulls as ingredients — Oat bran has been promoted heavily by large cereal companies, with the result that in early 1996, the FDA proposed allowing a product-specific health claim on food labels and in food labeling for that ingredient. Allowed claims would include calling attention to the alleged connection of consumption of oat bran or oatmeal (coupled with diets low in saturated fat and cholesterol) and reduced risk of certain kinds of circulatory diseases. Under the proposal, products for which such a claim is made must contain at least 20 grams of oatmeal or 13 grams of oat bran, and at least one gram of beta-glucan per serving.

Table 8.15
FORMULA FOR REDUCED-CALORIE BREAD WITH OAT BRAN

Ingredient	Control
Sponge	
Flour	60.
Bleached oat fiber	40.
Vital wheat gluten	5.
Compressed yeast	3.
Mineral yeast food	0.6
Ethoxylated monoglycerides	1.
Stearoyl-2-lactylates	0.5
Water	110.
Dough	
Flour	40.
High fructose corn syrup	7.
Vital wheat gluten	5.
Salt	3.
Compressed yeast	2.
Calcium propionate	0.5
Water	20.
Ascorbic acid	0.01

Figures are percents, flour weight basis. *Source*: Dougherty et al. (1988).

Small-scale preparation procedure

Mix sponge ingredients for one minute. Ferment for 3 hours at 86°F and 85% RH. Add dough ingredients to the fermented sponge and mix at low speed and medium speed, determining the appropriate duration by test. Rest mixed dough for ten minutes in fermentation cabinet. Divide into loaf size pieces, round, then rest 10 min. Form loaves on a cross-grain molder, and place in pans. Proof at 110°F and 81% RH for 60 minutes, then bake at 420°F for 18 min.

Results

The experimental bread had a fiber content of 4.1 grams per oz compared to 0.7 grams per oz for standard white pan bread. Corresponding figures for calories per ounce were 39.7 and 68.0.

Table 8.16

FORMULAS FOR SOFT COOKIES CONTAINING OAT FIBER AND OTHER FIBER SOURCES

Ingredient	Peanut butter cookies	Chocolate chip cookies
	Parts	Parts
First stage		
Oat fiber	29.	19.
Sugar beet fiber	2.	--
Cocoa fiber	--	12.
Unemulsified shortening	--	15.
Crunchy peanut butter	30.	--
High fructose corn syrup	70.	70.
Soy lecithin	--	2.
Second stage		
Flour	69.	69.
Dry whole eggs	1.4	1.4
Salt	0.9	0.9
Sodium bicarbonate	0.4	0.4
Barley malt	0.15	0.15
Natural peanut flavor	0.3	--
Water	40.	50.
Chocolate chips	--	--

Note: The shortening was Creamtex partially hydrogenated vegetable oil.
Source: Dougherty (1988)

Procedure

Mix in a bench mixer with a standard paddle. Blend first stage ingredients for three minutes at low speed. Add second stage ingredients for 1.5 minutes. Scrape bowl and paddle, then mix another 1.5 minutes. Rest dough ten minutes. Use a No. 40 disher to scoop out portions and deposit them onto silicone coated cookie sheets. Bake at 350°F for ten minutes.

Results

The peanut butter cookie contained 3.4 grams fiber per oz and the chocolate chip cookie 2.8 grams fiber per ounce as compared to an industry sample soft cookie without fiber additives containing 0.2 gram per oz. Calorie contents per ounce were 83.1, 83.6, and 133.6 (industry sample).

Bread made with the bleached oat hull preparation patented by Ramaswamy (1991) was described by the inventor. A sponge was made from 70% flour (these are bakers' percentages, i.e., based on total flour as 100%), water 57%, yeast 2.8%, and mineral yeast food 0.3%. After mixing, the sponge was allowed to ferment for four hours. Then, it was mixed with 30% four, 87% water, 2% yeast, 2.5% salt, 7% sugar, 2.5% milk solids, 3.6% shortening, 0.6% emulsifier, 35% oat fiber, and 14.8% vital wheat gluten. The oat hull fiber preparation was added after all the other ingredients had been well mixed. No floor time was allotted. Dough pieces were scaled at 550 grams and given an intermediate proof of 23 minutes, then molded. The loaves were proofed at 97°F for 39 minutes, then baked at 430°F for 29 minutes. The disclosure states that the bread had an excellent appearance in terms of break, shred, volume, and texture, and it had no off taste or gritty mouth feel. Unfortunately, the fiber content of the finished bread was not stated, but it probably was near 20% on an as is basis.

Psyllium seed — Psyllium seed husks have been suggested as a dietary fiber enhancement for bakery goods and other foods. This material has been used for a long time as the active ingredient of certain popular laxatives of the bulking type. It absorbs large amounts of water and is digested poorly, if at all, so that its caloric contribution is virtually zero. It has the advantage of being white and nearly tasteless, but it is rather expensive, and its very high water absorption qualities may cause difficulties in some applications. For an example of psyllium's use in foods, see the patent of Colliopoulos (1991).

Whole seeds have been used as a fiber source, but the special value of psyllium lies in the gum, sometimes called "hydrophilic mucilloid," that occurs in the seed coat. The gum can be readily removed from the seed coat by water extraction. Various species of plants are used as the source; among these are *Plantago lanceolate, P. rugeli,* and *P. ovata.* Commercial psyllium includes the French (black, *Plantago indica*) with about 11.8% gum content, Indian (blond) with about 30.9% gum, and German with about 11.5% gum.

Other concentrated fiber sources — "Fibrim" a trademarked soy fiber additive of Ralston Purina, is a food ingredient containing primarily the cell wall material of soybean cotyledon derived from processing dehulled and defatted soybean flakes. The manufacturer emphasizes that it is neither soybean hull nor soy bran. It is described as a bland, odorless powder available in several forms and contains about 75% total dietary fiber as analyzed by the AOAC method. Moisture 6.0%, protein 12.0%, Fat (ether extract) 0.2%, ash 4.5%, cellulose 10%, and hemicellulose and non-cellulosic polysaccharides 65%.

Oat soluble fiber β-glucan, is known to lower blood cholesterol in animals and humans. The fat substitute, "Oatrim," is derived from the bran or flour of oats by enzymatically converting the starch found in these raw materials. The degree of polymerization of the maltodextrins produced are important in determining functionality of the material. In addition, it is said that the soluble fibers bound in the cellular matrix are liberated. The three versions that have been offered are Oatrim-1 (from debranned whole oat flour), Oatrim-5 (from whole oat flour), and Oatrim-10 from oat bran. The numbers refer to the approximate percentage, on a dry basis, of the β-glucan present in the ingredient. Since the β-glucan is known to have cholesterol-lowering properties when consumed, a special dietary role is claimed for these fat replacing ingredients. The inventor says that many reduced fat and fat-free foods are possible using Oatrim for fat replacement. Oatrim is used broadly in many foods including low-fat meats, fat-free cheeses, and low-fat bakery products. Some sensory parameters of Oatrim usage in oatmeal raisin cookies and chocolate truffles were determined by an analytical sensory evaluation panel and reported by Inglett (1996).

Going back to basics, new varieties of grain with relatively high contents of fiber are being developed. Researchers at Montana State U. disclosed a new barley variety with enough beta-glucan to give it about twice the soluble fiber of regular barley. ConAgra is offering processed ingredients made from this grain to food manufacturers. Some of the forms are flakes, flour, and pearled.

BAKERY PRODUCT FORMULAS

It is not particularly difficult to formulate a food that is high in fiber and that meets regulatory requirements, but, when the added requirements of good organoleptic quality, satisfactory storage stability, and acceptable price are added, the problem becomes much more difficult. Bread-type products, including not only loaf bread and white rolls, but bagels, flatbreads, and other popular variations, have been the subject of much experimentation directed toward obtaining all-natural products deriving their fiber enhancement from whole grains (wheat and other cereals as well as non-cereal grains). Judging from the number of these varieties on the market, an economically viable demand continues to exist, though the texture and often the flavor of the breads leave much to be desired if the person is accustomed to the characteristics of fresh, well-prepared white bread and crusty rolls.

A relatively conservative use of multiple grains (as compared, say to the 10-grain breads that have been offered) is seen in the following formula and procedure developed by technicians at Cargill Foods. It is intended to be baked as hearth bread, in various sizes and shapes.

Formula

    7.35% cracked wheat cereal
  14.70% boiling water
    1.72% yeast (compressed)
  14.70% water
    1.72% vegetable oil
    1.35% molasses
    7.35% honey
  15.70% strong bread flour (Cargill Hummer brand)
  14.70% whole wheat flour
  10.78% rye flour
    6.25% fine semolina
    2.70% wheat bran
    0.98% sea salt

Process

Combine cereal with boiling water; set aside one hour to cool and absorb liquid. Dissolve yeast in remaining water, oil, molasses, and honey. Let stand ten minutes. Add softened cereal, flours, bran, and salt. Mix until a well-developed dough is obtained. Ferment until size is doubled, about 1½ hours. Punch. Ferment 30 minutes and punch again. Scale 20 oz pieces to yield baked loaves of one pound. Or, scale 7 oz for demi-loaves, 4 oz for rolls. Proof fully, about 45 minutes. Brush with an egg white wash then sprinkle with bran and slash the tops. Bake at 390°F until brown — typical times are: for loaves 25 to 30 minutes, for demis 17 to 20 minutes, and for rolls 15 minutes.

Results

This formula makes a very rough textured crumb, fairly soft but chewy (at least, when fresh). The fiber content that can be claimed will depend on the fiber in the specific ingredients used, but it would be expected to be better than ten percent on an as is basis.

Fiber-enriched Bakery Products

The dietary fiber content of bakery products can be increased by adding apple and pear fiber concentrates to the doughs and batters, generally replacing some of the flour, or by adding them on top of a ordinary formula. It is also said that these ingredients enable reductions to be made in the fat content of the formula because of their water-holding power. Apple pomace, the solid material left in the presses from cider manufacture, has received a good deal of attention for this application since the supply situation is good and the value of the raw material is essentially nil. Other fiber concentrates are also being offered for this purpose.

Fiber concentrates from field crops — Pie crusts in which fiber concentrates from field pea, selected soft wheat mill streams, soybean, sugar beet, and wheat residues were used to replace 5%, 15%, and 25% of the pastry flour were tested for acceptability by variously constituted panels of consumers. Ponte et al. (1995) stated that no significant differences in acceptability of the baked crusts were indicated, but the 5% fiber replacement and 35% water absorption samples made with soybean, wheat concentrate, and field pea rated higher than the others.

Campbell et al. (1994) described a set of experiments in which various constituents of an oatmeal cookie formula were modified in order to yield a high-fiber cookie of superior acceptability. One of the test formulas for this crisp cookie was (figures indicate parts, as is):

Formula — Part 1

Shortening, 10.00
Sugar, 6.50
Monoglycerides, 2.70
Water, 2.70
Polydextrose, 1.60
DATEM, 1.00

Formula — Part 2

Flour, 19.00
Powdered cellulose, 8.00
Oat fiber, 3.00

Formula — Part 3

Sugar, 6.50
Isomalt, 6.50
Polydextrose, 13.00

Formula — Part 4

Oats, ground, 13.00
Water, 9.00
Eggs, whole, 3.00
Other (flavors, etc.) 1.00

Conclusions of Campbell et al. included these: " ... of the three ingredient categories investigated, shortening substitutes had the greatest impact on measured texture attributes. Flour and sugar replacers, at the level used, had less or no significant effect on texture."

Bread with citrus pulp derivatives — The use of citrus vesicle cells in baked goods was disclosed in the patent of Patton et al. (1986), also a citrus product from orange and grapefruit waste containing the peels, membranes, seeds, and pulp combined with sesame seed flour. Following are the formulas and processes for examples disclosed in the publication. All proportions are given as percentages of the total flour weight, i.e., bakers' percentage. Procedures are summarized from the original text.

The following formula was used in a conventional straight dough processing scheme to yield acceptable loaf bread:

Formula

95% high protein spring wheat flour, about 14% protein.
5.375% yeast
115.6% water
0.5% Fermaloid (tm), an acid type mineral yeast food.
0.8% PD-121 (tm), a mixture of sodium and calcium stearoyl lactylate.
0.6% calcium sulfate
8% citrus pulp cell flour
5% gluten
15% oat flour
1.5% whey
2.875% salt
16.25% high fructose corn syrup
1.5% vegetable oil
0.4% calcium propionate
0.25% conditioner, mostly enzyme active soy flour and calcium peroxide.

Procedure
   Mix all ingredients at first speed for one minute, then at second speed for ten minutes, using a three-prong agitator. Bring outr at 78°F to 80°F. Bulk ferment for 1.75 hours. Punch the dough, then ferment for 30 minutes. Scale into 18.5 oz pieces, round, and rest 10 minutes. Sheet as thin as possible without tearing, and mold into loaf shape. Place in grease loaf pan. Proof 40 to 50 minutes in a chamber maintained at 110°F dry and 100°F wet readings. Bake for 30 minutes at 400°F. Cool for one hour, then slice.

Results
   In comparison with a white bread of conventional formula, the experimental bread contained 50 calories per oz vs 75 calories per oz, 46% moisture vs 38% moisture, 8% fiber vs 3% fiber, and 10% protein vs 8% protein. The experimental bread "has a similar appearance, feel, texture, and taste to conventional white bread."

   Patton et al. also provided a formulas and a process for making bread by the sponge process using citrus pulp to enhance the fiber content of the finished product.

Formula for sponge
   65% high protein spring wheat flour, about 14% protein
   5% gluten
   15% oat flour
   3% yeast
   114% water
   0.5% Fermaloid (tm), an acid type mineral yeast food
   0.6% calcium sulfate
   0.8% PD-121 (tm), a mixture of sodium and calcium stearoyl lactylate
   8% fruit fiber from citrus residue, etc.

Formula for dough
   30% high protein spring wheat flour, about 14% protein
   1.5% whey
   2.875% salt
   16.25% high fructose corn syrup
   1.5% vegetable oil
   0.25% calcium propionate
   0.25% conditioner, mostly enzyme active soy flour and calcium peroxide.
   112% water

Procedure
Thoroughly mix the sponge ingredients without developing. Ferment

for 3.5 hours, then mix with the dough ingredients. Ferment the dough for 30 minutes. Divide into 18.5 oz pieces, allow to rest for 10 minutes, then sheet, roll, and mold them into the size desired. Place the pieces in greased pans. Proof for 53 minutes in a chamber held at 110°F dry bulb and 100°F wet bulb. Bake at 400°F for 30 minutes.

Results

The baked loaf had "the taste, appearance, texture, and feel of conventional white bread." The experimental loaf has 10% protein content compared to 8% for conventional white bread, 46% moisture vs. 38%, 8% fiber vs. 3%, and 50 calories/oz compared to 75 calories/oz.

Cake mix with pineapple cellulose — A batter formulation for a reduced calorie cake using materials prepared from pineapple cores (a byproduct of pineapple canning operations) as a cellulose source was patented by Glicksman et al. (1985). Their description of an example follows (flavors, presumably vanilla, would have to be added, but are not given in the patent).

Formula

Cake flour, 35.92%
Sugar, 32.35%
Shortening, cake type, 7.19%
Egg white solids, 7.19%
Pineapple core bulking agent, 9.01%
Baking soda, 2.16%
Sodium aluminum phosphate, 2.16%
Nonfat milk solids, 1.82%
Sodium stearoyl-2-lactylate, 1.44%
Salt, 0.38%
Polyoxyethylene (20) sorbitan mono-oleate (Tween 80), 0.38%
    To each 100 parts of cake mix are added 112 parts of water.

Processing Method

First, a dry mix is prepared by usual methods of creaming the shortening, emulsifier, and sugar, then adding the remainder of the previously mixed ingredients and blending until uniform. Then water is added and mixed with a paddle agitator, first at low speed until the water is taken up, then at high speed for a total of about 2.5 minutes, with occasional scraping down of the mixer sides. Pour about 500 grams of batter into a greased 8-inch circular pan. Bake for 30 minutes at 350°F.

Results

There was a bake loss of about 10%. According to the inventors, the baked cake was of excellent overall quality and had a caloric value of 1.88 calories per gram as compared with conventional white cakes' 3.75 calories per gram. The moisture content of the cake was 47%.

Muffin mix with psyllium — Franssell and Palkert (1991) disclosed a microwavable muffin mix containing 51.8% flour, 24.2% sugar, 11.3% shortening, 4.3% dextrose, 3.1% banana flakes, 1.3% gelatinized starch, 1.0% toasted wheat germ, 1.0% salt, 0.9% sodium aluminum phosphate, 0.8% sodium bicarbonate, 0.2% banana flavor, and 0.1% caramel color, to which was added various amounts of psyllium and guar gum (e.g., 3.25% of each, on base mix basis). About 185 parts of the formulation were mixed with 90 parts of 2% fat milk and 50 parts of fresh whole eggs to make the batter. The muffins were said to have good texture, flavor, and appearance.

Pflaumer et al. (1990) obtained a patent on the use of psyllium seeds in cookies. One of the formulas they used as an example was as follows (amounts were in grams in the original publication):

Ingredients

 85.00 brown sugar
 26.28 water
  3.00 baking soda
  1.00 salt
120.00 shortening
 28.80 all-purpose flour
 14.00 pregelatinized starch, "Tendergel #434"
 61.20 psyllium seed coats, not finely ground
  3.72 dry egg white solids
  2.00 baking powder
 85.00 sucrose
110.00 peanut butter

Procedure

Mix the brown sugar, water, baking soda, and salt, then blend in the shortening. Flour, starch, psyllium, dried egg white, and baking powder are mixed together separately and then combined with the previously blend ingredients. Mix in the sucrose, then the peanut butter. Fifteen gram portions of the dough are placed on a metal sheet and baked at 375°F for about 13 minutes.

Results

The cookie had the taste and texture of a conventional peanut butter cookie. Six of the cookies per day are recommended as the proper dosage for reducing blood cholesterol or to treat gastrointestinal disorders.

The first example of Pflaumer et al. was modified in another formula they developed to further reduce the caloric content of the cookie by replacing the shortening with an equal amount of sucrose polyester. The cookie so obtained "had the texture and taste of a conventional [peanut butter] cookie."

Certain types of confections respond better to psyllium additions than are bakery foods. A type of imitation chocolate bar with a high content of psyllium was patented by Moskowitz (1988). An example given in the patent was: 230 grams of commercial confectionery coating (consisting of 29.5% fractionated palm kernel oil, 45% sugar, 12.5% cocoa powder, 0.1% vanillin, 0.4% soy lecithin, 7% whole milk powder, and 5.5% nonfat dry milk) is melted and 60 grams of whole psyllium husks are mixed into it; 20 grams of palm kernel oil is added and mixing continued until a uniform blend is obtained. The mass is cooled to about 36°C with occasional stirring and is then deposited into molds. After cooling to refrigerator temperature, the candy bar is demolded. The inventor states, "the products obtained have a crunchy mouthfeel and are highly palatable." They provide about 4.8 grams of dietary fiber per 1 oz serving. Other variables, mostly of flavor, are described in the patent.

Muffin Mix with barley β-glucan — Barley grains and fractions containing high levels of beta-glucans have been shown to have cholesterol-lowering effects in human. Hudson et al. (1992) described methods for making muffins of high fiber content from a barley fraction processed to have a particularly high level of the compounds of interest. The fraction was described as having 9.77% moisture, 12.81% protein, 3.71% fat, 3.37% fat, 80.11% carbonydrate, 42.08% total dietary fiber, 19.88% soluble dietary fiber, and 18.90% beta-glucan. Muffins were made using the formula and process described below (amounts in grams per laboratory batch).

Formula

Barley beta-glucan fraction, 127.2
Wheat flour, all purpose, 190.8
Baking powder, 8.0
Baking soda, 2.0
Salt, 1.2
Cinnamon, ground, 4.0
Buttermilk, liquid, 1% fat, 480.0
Egg white, 66.0
Safflower oil, 30.0

Honey, 120.0
Orange peel, grated, 8.0
Orange extract, 12.0 (ml)
Dates, chopped, 65.0

Process

Blended liquid ingredients were added to the combined dry ingredients and mixed by hand until just moistened, then the dates were folded into the batter. The batter was divided evenly into medium sized muffin cups which were then placed into muffin tins. Each muffin was topped with one gram of a 1:20 cinnamon sugar mixture and baked in a laboratory oven for approximately 20 minutes at 204°C.

Results

Three taste panels gave ratings of 6.7, 6.1, and 6.4 on a 9-point hedonic scale (9=like extremely, 5=neither like nor dislike, etc.). This compares to a rating of about 7.6 for a commercial oat bran mix. A 100 gram muffin contained about 188 calories and 13.88 total dietary fiber.

Methods for offsetting undesirable effects of fiber — It was contended by ter Haseborg and Himmelstein (1988) that the negative effects on bread loaf volume of hemicelluloses and pentosans could be offset by enzyme preparations containing hemicellulase or pentosanase. They conducted baking tests using the control formulas and procedures outlined in Table 8.17. Commercial preparations of the two types of enzymes were added at the rate of 15 grams per 100 pounds of flour, amounts that had been "established as optimum by preliminary qualitative tests." Enzyme supplementation caused significant volume increases in the test loaves, as compared to the control loaves. Doughs containing pentosanase additives were more pliable than controls, and doughs containing added hemicellulases were drier in appearance after mixing, with greater stability and less stickiness than the controls.

Sosulski and Wu (1988) prepared bread loaves by a straight dough method using a baker's patent flour made from hard red spring wheat and several fiber sources. They found it was possible to produce satisfactory bread using either 20% corn bran, 20% wheat bran, 15% field pea hulls, or 15% wild oat bran. Their formulas were optimized at 45% potassium bromate and 0.5% sodium stearoyl-2-lactylate; no supplemental vital wheat gluten was added. Prehydration of the fiber sources improved absorption and loaf quality. Results of analyses of the loaves for fiber content showed the following total dietary fiber: 21% for the corn bran loaves, 13% for the wheat bran loaves, 15% for the loaves containing pea hulls, and 6% for the loaves containing wild oat bran.

Table 8.17

FORMULAS AND PROCEDURES FOR HIGH FIBER BREADS

Ingredient parts or process time	Wheat bread	Whole wheat bread	Multi-grain bread	High fiber bread
Ingredients, parts				
Wheat flour	700	--	850	850
Water	670	670	650	1000
Whole wheat flour	300	1000	--	--
Microcryst. cellulose	--	--	--	220
7-grain mix	--	--	150	--
Sugar, granulated	--	--	--	120
Vital wheat gluten	30	30	30	110
Yeast	30	30	30	30
Shortening	30	30	30	15
Salt	20	20	20	20
Whey powder	--	--	--	20
Nonfat milk powder	20	20	20	20
Yeast food	10	5	5	5
DATA ester	--	--	--	5
Ascorbic acid	0.06	0.06	0.06	0.07
Potassium bromate	0.04	0.04	0.04	0.05
Process time, minutes				
Mix time at 70 rpm	3	3	3	3
Mix time at 140 rpm	3	3	4	5
Intermediate proof time	20	20	20	20
Final proof time	45	50	60	40
Baking time, 428°F	28	28	28	28

Source: Ter Haseborg, E. and Himmelstein, A. (1988), modified by the author.

Silva (1991) patented a method for improving the quality of bakery goods containing increased levels of dietary fiber by incorporating in, for example, wheat bran a sufficient amount of functionalized dairy products containing a thickening polymer. The thickener results from growing on a dairy product substrate (e.g., milk, defatted milk, whey, etc.) a polymer-generating microorganism from the genus Xanthomonas, e.g., *Xanthomonas campestris. Leuconostoc mesenteroides* can also be used. Details for preparing the thickening material can be found in the patent. A typical formula for high fiber bread made with this ingredient is given below (satisfactory but not optimal bread can be made with no shortening).

Formula
80 parts bread flour
20 parts wheat bran
67 parts water
 6 parts sugar
1.5 parts salt
 4 parts soy flour
3.5 parts bakers' yeast
 2 parts functionalized dairy product
 2 parts shortening
0.0088 parts ascorbic acid

Procedure
Blend all dry ingredients, then add the bakers' yeast and blend. Add the water (26.7°C) and mix in a jacketed bowl held at 23°C for one minute at low speed. Add shortening and continue to mix for 10 minutes, at medium speed. Bring dough out at 27.7°C. Hold the dough in a fermentation cabinet at 29.4°C for two hours. Scale out 625 gram portions of the dogh and hold for 10 minutes. Roll out the dough pieces, mold loaves, and place in pan. Proof at 43°C and 80% RH until the dough rises to about 1.25 cm above the edge of the pan. Bake at 196°C for 35 minutes.

Results
Loaves similar in texture and volume to conventional loaves were obtained.

Pasta Products

There does not seem to have been as much activity in devising spaghetti-type products of high fiber content as there has been in making bakery products enriched in this manner. Conventional pasta products consist almost entirely of durum semolina that has been mixed into a very dry dough, then extruded and dried. Consequently, the fiber content will be essentially that of the endosperm of durum wheat, which is not exceptionally high, being of the same order as white flour. There is no doubt a demand for high-fiber macaroni products. The following formula describes a pasta product said to have good extrusion properties as well as good organoleptic properties, while having a high content of fiber. The Solka-Floc ingredients are proprietary cellulose materials distributed by a division of Protein Technologies International (Anon. 1992B).

Formula

Durum flour, 47.7%
Solka-Floc 200 FCC, 14.0%
Solka-Floc 900 FCC, 1.0%
Vital wheat gluten, 2.0%
Guar gum, 0.2%
Egg flavor, 0.2%
Water, 20%
Whole eggs, 14.9%

Process

Pre-blend all dry ingredients. Add wet ingredients and mix. Extrude the dough through a standard macaroni press. Subsequently, treat as ordinary pasta.

Results

Extrusion rate (as compared to regular durum pasta) was increased by the cellulose additived. Water absorption was increasedin cooking. The final pasta product was rated between 5 and 8 on a 9-point hedonic scale. Calories of the cooked pasta were reduced by one-third over comparable serving weights of regular pasta. Total fiber was not given in this publication, but it appears to be about 20% on the material that comes out of the extruder, or about 25% to 27% on a dry weight basis.

Other Fiber Enriched Products

A patent by Garleb et al. (1992) discloses a formula and method for making a fiber mixture that can be used as an ingredient in a liquid food that constitutes part of weight-reducing regimen. The inventors considered several factors in designing this material, including what they considered as a correct proportion of non-fermentable and fermentable fiber, the correct proportion of soluble and fermentable (5% to 50%), soluble and non-fermentable (5% to 20%), and insoluble and non-fermentable (45% to 80%). the stability of the ingredient when made into a liquid food. The preferred system seems includes gum arabic, sodium carboxymethylcellulose, and oat hull fiber as well as various diluents, vitamin and mineral supplements, etc., to make up 100%. Examples of a complete powder system for reconstituting into a diet beverage included in the patent, but, since they include up to 80 ingredients, they will not be reproduced here. It is not known whether commercial products made according to this patent are currently being marketed.

An artificial rice disclosed in the patent of Hosoda, Hosoda, and Kato (1996) is compounded and formed to look like rice, but it has a poorly digestible polysaccharide as a main ingredient. It is made of an aqueous gel having as its main components glucomannan, starch, and dietary fiber. Although it is said to have an appearance and texture resembling cooked rice kernels, the inventors evidently expect it to be mixed with cooked rice in certain proportions, since they claim it does not separate from the natural material.

Breakfast cereals — In the Colliopoulas patent previously cited, we also find a formula for apple-cinnamon breafast cereal-type puffs. The formula is given below in terms of grams of ingredient per seven gram serving.

Psyllium, 3.500
Corn meal, yellow coarse, 1.085
Soy protein isolate, 0.840
Food starch, snack type, 0.700
Rice flour, 0.476
Natural apple flavor, 0.105
Oleoresin cinnamon, 0.105
Korintji cinnamon, ground, 0.091
Vegetable oil, partly hydrogenated, 0.049
Aspartame, 0.035
Lecitreme-3527 (Beatrice), 0.014

The manufacturing method was similar to that used for the apple-cinnamon flavored bar, i.e., it was hot extruded and cut to size. A different shape of die orifice was used, however, to make shapes like the common breakfast cereal puffs. Each seven gram puff is said to contribute about 13 calories in dietary energy, and it appears that the puffs would contain close to 50% dietary fiber, although this is not specifically stated in the patent disclosure.

High fiber bars and snacks — Several formulas for confection bars highly enriched with fiber can be found in the patent of Colliopoulos (1991). For example, he describes an apple-cinnamon flavored fiber bar containing a number of both insoluble and soluble fiber. The formula is as follows, amounts being kilograms of the ingredient (as is basis) in a thousand kilogram batch.

Formula

Refined corn bran G-fine, 157.50
Sized mucilloid, 155.00

Sucrose, 150.00
Fructose, 150.00
Yellow corn cones, 130.00
Soy protein isolate, 75.00
Rice flour, 50.00
Apple natural flavor, 25.00
Rice bran (Protex 40), 20.00
Oleoresin cinnamon, 20.00
Korintji cinnamon, ground, 20.00
Wheat bran, 17.50
Avicel pH 101, 15.00
Citrus pectin, 5.00
Guar gum powder, 5.00
Locust bean gum, 5.00

Procedure

(1) Blend all ingredients for about 15 to 20 minutes in a ribbon mixer.
(2) Extrude the mixture using a Creusot-Loire machine fitted with a centrol feed plate and brass die. Maintain a pressure of 1500 psi, screw speed of 150 rpm, and barrel temperature of 188°C. Add water at 150 ml per minute when the mixture is feeding at a rate of 1.686 kg per minute.
(3) The extrudate was cut into bars or wafers 9 grams in weight using a Leroy Savor Model MVS-50 cutter.

Results

The bars contained 3 grams of dietary fiber and had a caloric content of 24.

FEDERAL REGULATIONS AFFECTING FIBER CLAIMS

No development project for new dietetic products should be undertaken without a full knowledge of the effect of federal and state regulations on the labeling and marketing of the products. It is impossible to give in the limited space available here a complete review of the regulations, but a brief discussion of the most important points should be useful for the reader.

Labeling and Other Claims for Fiber

The FDA has stated, "Based on the totality of the publicly available scientific evidence, including recently available evidence, the agency has concluded that ther is no significant scientific agreement among qualified experts that a claim relating dietary fiber to reduced risk of cancer is supported. The publicly available evidence does indicate, however, that diets low in fat and rich in fiber-containing grain products, fruits and vegetables are associated with a decreased risk of several types of cancer, and there is significant scientific agreement that the evidence supports this association. The evidence is not sufficient to fully explain the role of total dietary fiber, fiber components and the multiple nutrients and other substances contained in these foods in reducing cancer risk."

So far as the effect of fiber on coronary heart disease is concerned, the FDA had this to say. "On the basis of the totality of the publicly available scientific evidence, including recently available evidence, the agency has concluded that there is significant scientific agreement among qualified experts that a claim relating diets low in saturated fats and cholesterol, and high in fruits, vegetable, and grain products that contain soluble fiber, to reduced risk of CHD is supported. The evidence is not sufficient to attribute the reduction in risk to soluble fiber or to a specific type or characteristic of soluble fiber, or to other components of these diets."

The risk factor used to assess the evidence was the presence of high levels of low-density lipoprotein in the blood. They considered whether the soluble fiber under investigation lowered low-density lipoprotein in a statistically significant and reproducible manner, without decreasing the amount of high-density lipoprotein (supposed to be advantageous).

Also, it was said, in referring to the results of studies comparing the effects of different sources of dietary fiber, "These results strongly suggest that the benefits of fiber supplements are not readily predictable by an analytical definition of soluble fiber but, rather, vary in some unknown way among different sources, or combination of sources, of soluble fiber. Thus, generalizing results from one fiber source to another must be done cautiously."

The FDA has approved health claims of the following general type for foods rich in dietary fiber (21 CFR 101.76):

1. Possibly beneficial in reducing the risk of cancer.

(a) Low-fat diets rich in fiber-containing grain products, fruits and vegetables may reduce the risk of some types of cancer, a disease associated with many [other?] factors.

(b) Development of cancer depends on many factors. Eating a diet low in fat and high in grain products, fruits and vegetables that contain dietary fiber may reduce your risk of some cancers.

2. For fruits, vegetables, and grain products that contain fiber, particularly soluble fiber, and that make claims about their effect on the risk of heart diseases, the following statements are suggested (21 CFR 101.77).

(a) Diets low in saturated fat and cholesterol and rich in fruits, vegetables, and grain products that contain some types of dietary fiber, particularly soluble fiber, may reduce the risk of heart disease, a disease associated with many factors.

(b) Development of heart disease depends on many factors. Eating a die low in saturated fat and cholesterol and high in fruits, vegetables, and grain products that contain fiber may lower blood cholesterol levels and reduce your risk of heart disease.

To justify the use of this health claim, the food must contain at least 0.6 gram of soluble fiber per serving. This limitation is based on the recommendations of an expert panel operating under the auspices of the Life Sciences Research Office, that the daily total dietary fiber intake should be 20 to 30 grams, and about 25% of this (roughly 6 grams) should be soluble fiber. The food must also meet requirements for "low saturated fat" (no more than 1 gram per serving), "low cholesterol" (no more than 20 mg per serving), and "low total fat" (no more than 3 grams per serving). Furthermore, the soluble fiber must be a natural constituent of the normal ingredients, not a high fiber supplement, such as some type of gum additive.

The current Daily Value of dietary fiber for labeling purposes is 25 grams. The NLEA regulations issued on December 2, 1992 require that the total dietary fiber calculation (which includes both soluble and insoluble fiber) no longer needs to subtracted from the Total Carbohydrate value and doesn't affect the calorie calculation. Only insoluble fiber, such as cellulose, can be subtracted from the Total Carbohydrate value to affect calorie reduction.

The FDA proposed a rule on January 4, 1996 that, if put into effect, would allow food marketers to claim that oat bran and oat meal reduce risks of coronary heart disease.

chapter nine

VITAMINS, MINERALS, AND PROTEINS

This chapter will include discussions of supplementation with vitamins, minerals, proteins, and amino acids — substances that can be added to foods in relatively small proportions so as to improve the nutritional value of the food for normal consumers. The avowed purposes of adding these nutrients are to (1) Replace substances lost in normal processing, as by discarding fractions containing desirable materials (e.g., in flour milling) or by heat destruction (in cooking or sterilization), (2) Improve the nutritional profile of foods that are naturally lacking in a substance needed by almost all normal consumers of that food (e.g., by adding vitamin D to milk), or (3) Make the food more suitable for specialized consumers, as by adding calcium to increase the food's value for older persons who might be susceptible to osteoporosis or iron for women in their middle years. Usually, it is not expected that the additive will replace a less desirable material or make the food more acceptable to persons suffering from some disease that requires a strictly controlled nutritional regimen.

REGULATIONS AFFECTING NUTRIENT ADDITIONS

Various rules and regulations issued by government agencies control the use of fortification and nutrient supplementation of foods, and limit the claims that can be made for such foods. These rules and regulations change from time to time, and it is advisable for all food manufacturers to keep themselves informed of the current status and prospective modifications of controls that will affect their products. Nutritional Quality Guidelines is a publication, or set of publications, that sets forth the general principles governing addition of supplemental nutrients to foods and their labeling; it may be available at your local library. New and revised editions and supplements are published from time to time, but there is, in fact, not much change in these guidelines. The laws themselves appear, eventually, in the Code of Federal Regulations (CFR), and are often summarized and explained in periodicals available to any manufacturer or food technologist.

There are also publications of the newsletter type that specialize in analyzing the legal import and significance of new and proposed regulations and reporting pending court cases and other legal actions pertinent to the food industry; these are often expensive. Much of this material is available not only in the paper versions, but also on-line, via the World Wide Web, and on CD-ROMs.

Federal Restrictions

Addition of nutrients to foods cannot be done solely at the discretion of the manufacturer. Certain federal regulations have been promulgated to prevent unnecessary or irrational supplementation. These are embodied in the regulations published in 21 CFR 104. A nutritional quality guideline prescribes the minimum level or range of nutrient composition appropriate for a given class of food.

Labeling for a product that complies with all of the requirements of the nutritional quality guideline established for its class of food may state, "This product provides nutrients in amounts appropriate for this class of food as determined by the U.S. government." The words "this product" can be omitted, if desired. If the manufacturer uses this statement, it must be printed on the principal display panel, and may also be printed on the information panel, in letters not larger than twice the size of the minimum type required for the declaration of net quantity of contents. Labeling of noncomplying products may not include this statement or in any other way imply that the item is in compliance with a guideline.

A product that bears the standard statement given in the preceding paragraph must also meet the following requirements: (1) The label of the product must bear nutrition labeling in the standard format as well as all other labeling declarations prescribed by the regulations, and (2) The label must bear the common or usual name of the food.

No claim or statement may be made on the label representing, suggesting, or implying any nutritional or other differences between a product to which nutrient addition has or has not been made in order to meet a guideline, except that nutrient additions shall be declared in the ingredient statement.

A product in a class of food for which a nutritional quality guideline has been established and to which has been added a discrete nutrient either for which no minimum nutrient level or other allowance has been established as appropriate in the nutritional quality guideline, or at a level that exceeds any maximum established as appropriate in the guideline, shall be ineligible to bear the guideline statement. Such a product shall also be deemed to be misbranded under the act, unless all labeling bears conspicuously the statement, "The addition of [added nutrient] to this product has been determined by the U.S. Government to be unnecessary and inappropriate and does not increase the dietary value of the food." Optionally, the words, "The addition of [added nutrient] at the level contained in" may be substituted for the first six words of the preceding statement.

Regulating Nutrient Additions

This part of the regulations establishes a uniform set of principles that can serve as a model for the rational addition of nutrients to foods. The achievement and maintenance of a desirable level of nutritional quality in the nation's food supply has been accepted as an important health objective. The addition of nutrients to specific foods is said to be an effective way of maintaining and improving the overall nutritional quality of the food supply. Random fortification of foods could, however, result in over- or under-fortification in consumer diets and create nutrient imbalances. It could also result in deceptive and misleading claims. It was recognized that the policy set forth may not be applicable to conditions existing in other countries.

Table 9.1

REFERENCE DAILY INTAKES FOR VITAMINS AND MINERALS

Nutrient and unit	Daily value
Vitamin A, IU	5000
Vitamin C, mg	60
Calcium, g	1
Iron, mg	18
Vitamin D, IU	400
Vitamin E, IU	30
Thiamin, mg	1.5
Riboflavin, mg	1.7
Niacin, mg	20
Vitamin B_6, mg	2.0
Folate, mg	0.4
Vitamin B_{12}, µg	6.0
Biotin, mg	0.3
Pantothenic acid, mg	10
Phosphorus, g	1
Iodine, µg	150
Magnesium, mg	400
Zinc, mg	15
Copper, mg	2.0

Source Browne (1993).

Fortification — The FDA does not encourage indiscriminate addition of nutrients to foods, nor does it consider it appropriate to fortify fresh produce, meats, sugar, or snack foods such as candies and carbonated beverages. To preserve a balance of nutrients in the diet, manufacturers who elect to fortify foods are urged to utilize these principles when adding nutrients.

The nutrients shown in Table 9.1 may appropriately be added to a food *to correct a dietary insufficiency* recognized by the scientific community to exist and known to result in nutritional deficiency disease if:

(1) Sufficient information is available to identify the nutritional problem and the affected population groups, and the food can act as a vehicle for the added nutrients. Manufacturers who wish to use this principle should contact the FDA before implementing a fortification plan based on this principle.

(2) The food is not the subject of any other Federal regulation (such as a Standard of Identity) for a food that requires, permits, or prohibits nutrient additions.

Restoration — A nutrient listed in the table may appropriately be added to a food to *restore its content* to a level representative of the food prior to storage, handling, and processing, in cases where:

(1) The nutrient is shown by adequate scientific documentation to have been lost in storage, handling, or processing in a measurable amount equal to at least 2% of the U.S. RDA (and 2% of 2.5 grams of potassium and 4.0 mg of manganese, when appropriate) in a normal serving of the food.

(2) Good manufacturing practices and normal storage and handling procedures cannot prevent loss of the nutrient.

(3) All nutrients, including protein, iodine, and vitamin D, that are lost in a measurable amount are restored and all ingredients of the food product that contribute nutrients are considered in determining the restoration levels.

(4) The food is not the subject of any other Federal regulation that requires or prohibits nutrient additions, or the food has not been fortified in accordance with any other Federal regulation that permits voluntary nutrient additions.

When enriched in accordance with the preceding paragraph, a food may be labeled "fully restored with vitamins and minerals" or "fully restored with vitamins and minerals to the level of unprocessed [common or usual name]."

In 1996 the FDA deleted its prohibition on addition of vitamins to peanut butter so as to allow manufacture of modified products (such as reduced fat peanut butter) that would be nutritionally equivalent to standard peanut butter (61 Federal Register 7207 et seq.).

Supplementation — A nutrient listed in the table may be added to a food in proportion to the total caloric content of the food, in order to *balance the vitamin, mineral and protein content*, if the following requirements are met:

(1) A normal serving of the food contains at least 40 kcal (i.e., 2% of a daily intake of 2,000 kcal).

(2) The food is not the subject of any Federal regulation for a food that requires, permits, or prohibits nutrient additions.

(3) The food contains all of the nutrients shown in the table per 100 kcal based on a 2,000 kcal total intake as a daily standard.

When enriched in accordance with the preceding paragraph, the food may include on its label the claim, "vitamins and minerals added are in proportion to caloric content." The word "protein" may be included, if appropriate.

Replacements — To avoid nutritional inferiority, a nutrient may appropriately be added to a food that replaces traditional food in the diet provided the substitute food is not nutritionally inferior to the traditional food and the product bears on the label a common or usual name that is not false or misleading — if there is no common or usual name, an appropriately descriptive term that is not false or misleading may be used. For such foods, it is not appropriate to make a label claim that vitamins or minerals have been added. Of course, the additives must be included in the ingredient statement.

Other restrictions — A nutrient addition is appropriate only when the additive is stable in the food under customary conditions of storage, distribution, and use. Also, it must be physiologically available from the food, and must be present at a level that gives a reasonable assurance that excessive amounts of the nutrient will not be consumed in normal diets. Additionally, the nutrient must be suitable for its intended purposes and be in compliance with applicable provisions of the act and regulations governing the safety of substances in food.

When labeling claims are permitted, the terms "enriched," "fortified," "added," or similar terms may be used interchangeably to indicate the addition of protein or one or more vitamins or minerals to a food, unless an applicable Federal regulation requires the use of specific words or statements.

Legal status of additives — Vitamins are GRAS (generally recognized as safe) and can be added to nonstandardized bakery foods. Minerals are always, or nearly always, added in the forms of their salts (e.g., calcium as calcium carbonate) and some of these salts are also GRAS.

An excess of certain vitamins can have harmful effects on consumers, and people vary in their susceptibility depending on their age, health, etc. It is highly unlikely that a harmful excess of any vitamin or mineral could be ingested as the result of overindulgence in bakery products, even if the products were supplemented at fairly high levels.

Proteins and amino acids are not necessarily GRAS, especially in the forms that would be available for food enrichment. Each candidate must be separately reviewed. Exceptions might be the use of a GRAS ingredient that is naturally high in protein, such as egg white or nonfat dry milk.

Table 9.2
FDA STANDARDS FOR ENRICHED FLOUR AND BREAD EFFECTIVE JANUARY 1, 1998[1]

	Enriched flour	Enriched bread
Thiamin	2.90	1.80
Riboflavin	1.80	1.10
Niacin	24.00	15.00
Iron	20.00	12.50
Folic acid	0.70	0.43
Calcium (optional)	960.	600.

[1] Expressed as milligrams of nutrient per pound of product.
The flour standards are also applicable to enriched self-rising flour and enriched bromated flour.

ENRICHMENT WITH MINERALS

The class of micro-nutrients can be considered as including vitamins and essential minerals. Few, if any, natural foods contain a full complement of all these substances. This is the reason that nutritionists advocate a diet containing a number of different foods. However, foods that are most liked for their taste, texture, and appearance, or are consumed as a result of ethnic or traditional customs, often have very low amounts of vitamins and essential minerals. For these reasons, it has become the practice to add small amounts of concentrated vitamin and mineral sources to foods that would otherwise be nutritionally insignicant.

Bioavailability

A major problem in the addition of micro-nutrients is physiological availability or bioavailability of the compounds added. This difficulty is more common in mineral addition than in vitamin supplementation. Even though a chemical or physical analysis can detect, say 300 mg of calcium in a portion of food, this does not indicate that the person who consumes that food will have all (or, indeed, any significant amount) of that 300 mg of calcium enter his bloodstream.

Solubility is one of the major factors affecting physiological utilization of mineral supplements, but the ability of the body to extract the needed element or mineral from more complex compounds must also be considered. Binding of otherwise soluble micro-nutrients by dietetic components such as insoluble fiber has been recognized as a significant problem. Will consumer treatment of the product, as in preparation of a cake mix, cause loss of nutrients?

Other Considerations

The economic impact of nutrient addition is also a factor that enters the equation; not only must the actual cost of ingredients (often very expensive on a per pound basis) be considered, but also the cost resulting from necessary changes in processing, packaging, and labeling. Reduction in shelf-life may be a problem.

Formulators must keep in mind the possible interaction of added nutrients. It is well-known that vitamin C will improve the absorption of iron, but iron can also accelerate vitamin degradation in food. This degradation can lead to changes affecting the acceptability of the product, such as off-color development. Copper can degrade vitamin C.

The minerals most commonly added to bakery products are calcium, phosphorus, and iron. Iodine and magnesium are sometimes added. The other metallic elements are rarely added.

Minerals such as iron and copper can cause unwanted changes in the flavor and color of foods. The form of the mineral selected for fortification must be considered with this possibility in mind. Calcium, in particular, can cause changes in the texture or mouthfeel of products and the response to processing conditions (such as mixing) of intermediates, especially if the calcium compound selected as an enrichement is soluble.

Calcium and Phosphorus Enrichment

Because calcium and phosphorus are frequently added in one compound, as some type of calcium phosphate, and since the physiological effects of the two elements are intertwined, it is sensible to consider them together. It is also apparent that the physiological actions of the two elements are, in some respects, connected.

Calcium is an essential nutrient. Americans get the largest part of their calcium intake (about 55% on average) from dairy foods; the remainder comes mostly from leafy vegetables and calcium-fortified foods. The average daily intake varies from about 300 to about 1,200 mg of calcium. Many bakery products contain enough milk solids to give a significant boost to the the product's calcium and phosphorus content. Several other food ingredients can be used to increase the calcium and/or the phosphate content of foods.

Role of calcium in health and illness — Calcium plays many roles in human metabolism. Bones, the supporting structure of which is largely a type of calcium phosphate, are dependent on a supply of calcium ions for their formation and continued integrity, but there are many other functions of calcium in the human organism, some of them essential to the most fundamental reactions. For example, dietary calcium plays some role, not clearly defined, in controlling blood pressure. Bucher et al. (1996) suggest a hypothesis that "inadequate calcium intake is associated with increased blood pressure that can be corrected with calcium supplementation." These authors reviewed a number of other studies, and found that calcium supplementation may lead to a small reduction in systolic but not diastolic blood pressure. This does not seem to provide clear-cut evidence for their hypothesis, but they said, "The results do not exclude a larger, important effect of calcium on blood pressure in subpopulations. In particular, further studies should address the hypothesis that inadequate calcium intake is associated with increased blood pressure that can be corrected with calcium supplementation."

McCarron and Hatton (1996) say, " . . . meta-analyses of randomized controlled trials of blood pressure and calcium levels in 2,412 adults and in 2,459 pregnant women provide compelling evidence that both normotensive and hypertensive individuals may experience reductions in blood pressure when calcium intake is increased."

Elderly persons are particulary interested in calcium supplementation because of the apparent role of dietary calcium in supporting bone growth and retention. Each year, about 4% of the population of older women sustain hip fractures, and there are other serious problems with bone degeneration, distortion, and weakening of the skeleton with advancing age. Decreased calcium intake resulting from changes in food selection has long been advanced as a possible cause of these changes, but recent studies suggest that low intake of Vitamin D is implicated. Apparently, low availability of the vitamin results in noncompensated increases in parathyroid hormone, producing negative calcium balance and a gradual loss of bone mass. The picture is further complicated because of reduced exposure of the senior citizens to sunlight, with a resultant decrease in the amount of vitamin D generated in the skin.

Calcium and phosphorus supplements — Phosphates of calcium phosphate should be considered as fortification compounds if both calcium and phosphorus are to be added to the food. Economic, organoleptic, and functional reasons favor these compounds, but none of the readily obtainable calcium phosphates contain nutritionally balanced amounts of the two nutrients, so that it is not possible to add a single compound that contains just the minimum amount of each. If only calcium is to be added, calcium carbonate, calcium sulfate, and calcium oxide are possibilities — each has some negative features that should be fully explored before making a final decision.

Nonfat dry milk and/or dried whey are useful "natural" calcium supplements, but they are very expensive compared to the inorganic forms, especially if the other functional values supplied by these materials are not needed in the product. Flavor, color, and processing difficulties may ensue if high levels of these milk derivatives are added to supply adequate amounts of calcium and/or phosphorus.

The bioavailability of calcium in each possible choice of supplement must be considered. It is not desirable to claim that a supplement adds 100 mg of calcium (for example) to the diet when the compound being added is totally insoluble and its calcium content passes through the intestine without being absorbed. Unfortunately, measures of calcium bioavailability are not easily obtained, and, when obtained, may not be applicable to use of the supplement in all foods and for all persons. When the bioavailabilities of calcium in six different sources (whole milk, chocolate milk, yoghurt, imitation milk, cheese, and calcium carbonate) were compared, the mean absorption values of all sources were clustered between 21% and 26% and none differed significantly from the others (Recker 1988). Bioavailability of calcium from other sources, particularly from some inorganic phosphates, may not be as good.

Calcium and phosphorus are frequently added in the form of a relatively inert (chemically) variety of calcium phosphate. Since these salts must be added in large amounts to achieve supplementation at a substantial percentage of the recommended daily amount per portion, undesirable effects on doughs (or other intermediates) may be observed. Some calcium phosphates have high buffering capacity and may cause a change in the intermediate's pH. Calcium ions can have a toughening or insolubilizing effect on gluten, thus changing the machining behavior of the dough. The finished product often exhibits an improved loaf volume and better grain and texture. No change in flavor need occur if the compound is properly chosen. Calcium also interacts with pectic acids to affect texture in foods in which these materials cause gelling.

The largest fast food operator requires its bakery suppliers to enrich with calcium the hamburger buns it buys, reports Jackel (1987). This author further states, some wholesale bakers (but not all of them) have been enriching white bread and rolls to the maximum allowable level of 600 mg per pound since nutritional labeling became effective. Many other wholesale bakers have not formulated to the maximum level, but have relied on calcium contributions from the various calcium-containing ingredients normally used in the formulation.

Calcium sulfate, which can be obtained from large natural deposits, has become accepted as the major calcium supplement in baking because it is economical in price, is well absorbed by the human digestive tract, and generally has no bad effects on baking quality (according to Jackel). Calcium can be added in the bakery as a separate ingredient or it can be added in the form of calcium-enriched flour containing 960 mg of calcium per pound. To increase to 600 mg per pound the amount of calcium in white bread of typical formulation, it is necessary to add about 0.375% calcium sulfate dihydrate based on flour weight as 100%.

The variety of calcium sulfate chosen by nearly all bakers who use the sulfate form of calcium enrichment is the dihydrate. Although anhydrous calcium sulfate can be used satisfactorily, it is somewhat more expensive than the dihydrate and is not known to function any better. The supplement can be added as a dry ingredient to the dough, or in some cases, to the sponge. In pre-ferment or broth systems, the calcium sulfate can be added to the brews or, later, in the ingredient premixer.

Jackel claims that all calcium enrichments except calcium sulfate can cause problems. Calcium carbonate and monocalcium phosphate interfere with fermentation and change the dough pH. Calcium propionate introduces flavor effects and interferes with fermentation, greatly slowing proof time. Using dairy products as a source of calcium would require addition at such excessively high levels they would cause reduced loaf volumes and coarse grain and texture.

The limit of 600 mg of calcium per pound of finished product, previously mentioned, applies to Standard of Identity bread, rolls, and buns. If the product is not a Standard of Identity item, i.e., not called white bread or another of the names prescribed, perhaps a "Farmer's Hearth Loaf" or other type not purported to be a white bread, higher levels of supplementation are allowed. Jackel mentioned a "Special Formula" bread, first marketed in early 1986, that carries a calcium level of 135 mg per serving. This allows it to indicate on the label that a serving (in this case, 18 grams) contains 15% of the US RDA, and to claim that it is a "significant source" of calcium. The added calcium does not interfere with body, strength, grain, or texture of the thin (5/16th inch) slices of bread, and in some cases appeared to strengthen the dough and improve the internal characteristics of the baked product. It was known that, in Standard of Identity white bread, a chalky flavor can be detected when the amount of calcium is increased by 50% to 100% over the allowable 600 mg of calcium per pound of bread. Test results showed that adding molasses, malt syrup, and honey to the formula allowed the amount of calcium sulfate dihydrate to be increased more than five-fold without the development of a chalky flavor and mouthfeel.

The large amounts of calcium and phosphorus needed to supply meaningful percentages of the US RDA to the consumer can lead to seemingly insuperable problems in formulating certain types of highly enriched baked products. If only 10% of the US RDA of phosphorus and calcium is the intended supplementation, there would have to be added about 0.5 gm of tricalcium phosphate per portion of food. Textural and visual defects can result from the inclusion of such quantities of phosphate salts to portion quantities of most cookies or crackers, and perhaps of bread and rolls.

Calcium supplementation is not always a matter of adding inorganic salts of the element. Calcium lactate, calcium gluconate, calcium citrate, and possibly other calcium salts of organic acids are available for use as nutrient supplements. Because they are considerably more expensive than calcium sulfate, calcium carbonate, or the phosphates of calcium, their applications tend to be restricted to specialized categories of foods where their special advantages are critical to satisfactory performance. The content of the active ingredient (calcium) is also a consideration. Calcium fumarate contains 25.9% calcium, calcium aspartate 23.4%, tri-calcium citrate tetrahydrate 21.0%, calcium lactate trihydrate 14.7%, and calcium gluconate 9.3%.

Calcium lactate is a highly soluble (about 9 grams of the salt will dissolve in 100 grams of water) calcium salt that is primarily promoted as a fortifier for beverages. The "natural" form is produced by fermentation of sugar and is based on the L(+) isomer of lactic acid, the type found in the human body. It is more expensive, per unit of calcium, than most of the phosphorus salts, but it could be valuable in food supplementation in cases where phosphates are contraindicated or where higher solubility and bioavailability are desired (Labin-Goldscher and Edelstein 1996).

Fruit drinks pose special problems when calcium supplementation is being considered. Direct addition of calcium sources such as calcium carbonate or calcium hydroxide to orange juice can cause undesirable cooked or browned off-flavors and may contribute to the stripping of desirable flavor and aroma compounds. Adding calcium chloride at levels above about 0.11% can impart a brackish flavor to the juice, an it reduces the normal flavor intensity and quality.

The patent of Heckert (1988) reveals the use of calcium citrate and calcium malate, supplements that are said to avoid the usual disadvantages of adding substantial quantities of other calcium salts to fruit juices. Usually, these two calcium salts are formed in the beverage from precursors. An example given in the patent, resulting in an apple juice beverage containing 0.12% calcium is:

243.22 g apple juice concentrate of 70° Brix
212.85 g apple aroma concentrate
  4.26 g calcium concentrate
  1.06 g citric acid
  2.98 g malic acid
954.63 g water

The process described in the Heckert patent, includes the following steps: Make a premix solution by dissolving the acids in 400 g of the water, then carefully add the calcium carbonate with stirring. After foaming has ceased, add the premix solution to the apple concentrate. Finally, add the apple aroma concentrate and the remaining amount of water. Stir the batch vigorously and bottle.

The body's absorption of calcium from the intestinal contents can be interfered with by other food components. Among these are oxalic acid found in some greens, rhubarb, and peanuts, phytic acid, found in the outer portions of grains, and insoluble fiber, found in many foods. Lactose, or milk sugar, increases calcium absorption, unless the subject is lactose-intolerant, in which case calcium absorption is hindered.

Iron

Iron as a food additive has both positive and negative aspects. Several nutritional surveys have indicated that as much as 20% of the U.S. population may have "unacceptably" low levels of iron in their blood. On the other hand, some people react adversely to relatively small increases in iron intake. From time to time, various groups have pressured the goverment to increase the allowable iron supplementation in common foodstuffs, while other groups have emphasized the possible ill effects such action might have on a substantial part of the population. Since the early 1940s, the FDA has required the addition of iron to wheat flour, farina, bread, buns, and rolls. In 1973, increases in iron to 40 mg per pound of enriched flour and to 25 mg per pound of enriched bread, buns, and rolls were approved. These increases were rescinded in 1977. Effective July 1, 1983, single-level requirements of 12.5 mg per pound for enriched bread, rolls, and buns, and 20 mg per pound for enriched flour were established.

Biovailability of iron in common foods — It has been found that different chemical forms of iron are utilized to different extents by the human body. Thus, 1 mg of iron in a certain compound may pass through the intestinal tract practically unchanged and unabsorbed, while 1 mg of iron in the form of another compound may be almost totally transferred to the circulating blood for utilization in body processes. The food value or absorption of dietary iron is affected more by the chemical state of iron in the food than by the total amount of iron.

The source and chemical form of dietary iron markedly affect the nutrient's availability for absorption. Morck and Cook (1981) point out that a typical Western diet will contain four sources of iron: food of animal origin, food of vegetable origin, fortification iron, and contamination iron. For much of the world's population, animal foods are in short supply, and fortification iron is probably seldom encountered. Because of limitations of space, and taking into account the probable major interests of readers of this book, only Western dietary practices will be considered in the following discussion.

The two major chemical forms of iron in a mixed diet (i.e., not strictly vegetarian, or otherwise deliberately or unintentionally restricted) are heme iron and nonheme iron.

Heme iron provides about 5% to 10% of the total iron ingested in a typical mixed diet, but because of its high bioavailability it furnishes about a third of the iron absorbed each day. It is found in the hemoglobin and myoglobin that are present in animal tissue (including fish and poultry), where it is bound in a porphyrin ring structure that is more soluble in the neutral conditions existing in the small intestine than it is in the acidic environment of the stomach. Heme iron is absorbed by mucosal cells as the intact metalloprotein, and the iron is split from the protein when it gets inside the cells. This is a different mechanism than prevails for the remainder of dietary iron, and, as a result, heme iron biovailability is unaffected by dietary components that either enhance or inhibit the availability of nonheme iron. Absorption of heme iron averages about 15% for iron-replete men to 35% for persons deficient in stored iron.

Vegetable foods do not contain heme iron but both animal and vegetable tissues contain nonheme iron. For this form of iron to be absorbed into the human system, it must reach the mucosa of the upper small intestine in a soluble form. Its absorption can be either aided or hindered by other dietary constituents.

Food additives for iron supplementation — Ingredients added to foods to increase their dietary iron content always, or almost always, take the form of nonheme iron. Several different iron compounds, as well as elemental iron, have been offered as additives for enriched flour and bread. The most common enrichment substances are either soluble iron salts or elemental iron in very small particles. When ingested without accompanying food, ferrous salts are absorbed better than ferric compounds. Although the particle size of elemental iron is known to have a significant influence on the substance's solubility and availability for absorption, electrolytic iron has a higher bioavailability than hydrogen-reduced iron of the same particle size, possibly due to the greater surface area of the former material.

Different iron compounds differ in the availability of their iron content for nutrition of animals, including humans. Some of them are poorly absorbed and may pass out of the digestive tract unchanged, while others are taken into the bloodstream fairly rapidly and completely. The bioavailability of iron compounds is affected by other materials in the diet, e.g., calcium, phytates, and ascorbic acid. Bioavailability also varies according to the condition of the consumer's metabolism. Iron from animal sources is almost always more readily available than iron from plant sources, evidently due to the heme structures in the former. Of the iron compounds that have been suggested for use as nutrient additives, ferrous sulfate, ferric ammonium citrate, ferrous fumarate, and ferrous gluconate seem to have relatively high bioavailabilities. Some forms of elemental iron also have rather good availabilities, while sodium iron pyrophosphate and ferric orthophosphate are regarded as having low bioavailability, at least under some circumstances.

According to Morck and Cook (1981), the very small amount of iron absorbed relative to the iron content of the diet makes chemical iron balance studies in humans very difficult. Use of radioactive iron or biosynthetically tagged heme iron facilitates such studies. Analytical or sampling errors may mask the effects on iron balance attributed to the diet. Cook et al. (1973) showed that solubility alone cannot accurately predict bioavailability of different iron compounds used for fortification of wheat flour. They used four soluble forms of iron as ingredients in dinner rolls, and measured the rate of iron absorption in humans. Reduced iron was absorbed the same as ferrous sulfate, but ferric orthophosphate and sodium iron pyrophosphate showed only about 30% and 50% availability, respectively.

Mineral supplements can be obtained from several manufacturers. The different enriching compounds vary widely in cost per pound of effective ingredient (iron). They can be ordered in bulk containers, or arrangements can be made with the supplier to furnish them in preweighed packets containing sufficient nutrients to bring one batch up to specification levels. Tablets containing vitamin and mineral enrichment in the portions required for one batch of enriched bread dough are also available. And, of course, enriched flour is available from nearly every wheat miller.

Effect of other food components on iron availability — Components of foods consumed at the same time, or at about the same time, as an iron supplement can affect the bioavailability of the element, in either a positive or negative direction. Although an iron salt may exhibit a high level of availability when it is fed separately from other dietary components, if it is added to a meal of corn or wheat (for example), the element's bioavailability may be reduced considerably. On the other hand, iron absorption can be markedly increased if the nutrient is accompanied by a source of ascorbic acid.

The interaction of iron with other dietary and with compounds resulting from enzymatic digestion greatly complicates the determination of bioavailabilty of dietary iron supplements. For example, the presence of heme iron actually increases the absorption of nonheme iron, while coffee and tea substantially decrease the absorption of iron.

Calcium and phosphate compounds have been reported to decrease the availability of nonheme iron. Both the white and yolk of eggs can inhibit the absorption of nonheme iron contributed by other foods eaten in the same meal. Iron absorption is reduced in direct proportion to the amount of wheat bran added to a meal, possibly due to the high fiber content of this grain constituent. Pectin can also reduce iron absorption.

In a study comparing the bioavailability of iron in white bread fortified with ferrous sulfate, ferric sodium pyrophosphate, and electrolytic iron at two levels, it was found that iron distribution was affected by the amount and types of other ingredients in bread. Besides the initial form of the iron, the ingredients that had the most effect on the soluble iron when it was in an aqueous slurry were shortening, salt, and nonfat dry milk. When mixed with similated gastric juices, the bread ingredients sugar and shortening had the most effect. Under simulated duodenum conditions, soluble iron was decreased by high salt levels and increased by high yeast levels. The authors (Kadan and Ziegler 1986), believe that the inhibitory effect of salt on iron bioavailability is unique and important.

Adverse effects of added iron — Iron salts can cause problems related to undesirable changes in organoleptic properties. Soluble iron salts, such as ferrous sulfate, can react with cocoa and perhaps with other foods to produce blue, gray, or black hues which are quite objectionable in many foods. Reduced iron may be observed as small black specks, and because of its high density it separates from the lighter ingredients and, if it is in a dry mix, it tends to move toward the bottom of the contain during handling and transportation.

Some iron enrichments greatly accelerate the development of rancidity in cookies or other fat-containing foods. Sodium iron pyrophosphate is less troublesome in this respect and has a number of other advantages over alternative iron supplements, being bland in flavor, white in color, and relatively insoluble. Its main disadvantages are a relatively high cost per unit of iron supplied to the consumer and an apparently low level of bioavailability.

Processing of foods can change the bioavailability of iron present in the ingredient mix. In 1980, Lee and Clydesdale reported that the baking process generates large amounts of insoluble iron, regardless of the iron source added to the baked food. In addition, pre-baking differences between iron sources were eliminated by the baking process. These results suggest that baking can affect iron absorption.

GRAS status of some iron compounds — In 1988, the FDA affirmed GRAS status for elemental iron, the ferrous salts of ascorbate, carbonate, citrate, fumarate, gluconate, lactate, and sulfate, and ferric salts of ammonium citrate, chloride, citrate, phosphate, pyrophosphate, and sulfate. Other compounds remaining under consideration until further information is received are ferric sodium pyrophosphate, iron peptonate, iron polyvinylpyrrolidone, sodium ferric ethylenediamine tetraacetate, and sodium ferricitropyrophosphate.

Other Minerals

Iodine is usually added as potassium iodide or, for convenience of measuring and mixing, in the form of iodized salt. The amount required to meet enrichment standards is so small that the cost is almost negligible, even for fairly high levels of supplementation. Effects of iodine on color, flavor, texture, and functional properties should be undetectable at all normal addition levels.

If it is necessary to add magnesium to bakery products, magnesium chloride is considered the compound of choice. The oxide, another cheap source of the element, may have some effect on pH of the dough, but at ordinary and acceptable levels of addition, the magnesium ion probably has little effect on either the physical properties of doughs and batters or the characteristics of finished products.

Diets deficient in manganese may lead to osteoporosis, according to recent reports. If the diet is also high in calcium, the problem may actually be accentuated, contrary to expectations. The estimated safe and adequate level of dietary manganese is 2.5 to 5 mg per day, but some authorities say this estimate may be too low. One study showed that menopausal osteoporotic women had an average serum manganese level only 25% that of age-matched norml women, a tentative indication that manganese plays a key role in bone development. Wheat bran and certain other plant derivatives are fairly high in manganese, but in them the element is present in forms that are almost totally resistant to digestion.

Although boron is not considered an essential nutrient, a USDA study indicates that boron may play a very important role in strengthening bones (Anon. 1988A). This work showed that calcium and magnesium losses in postmenopausal women were lower when they consumed boron-rich foods than when they were on normal diets. Boron is adequately supplied by most vegetarian diets.

Storage losses

Since mineral supplements are not volatile and do not normally undergo storage reactions that destroy the active constituent, the percentage of the element placed in the food at the time of its manufacture should be the same as that reaching the consumer. That is, storage losses need not be considered in calculating the amount of supplement to be added. Some questions have been raised about the possibility of certain minerals reacting with other ingredients or atmospheric gases to form compounds that cannot be utilized by the body, or reacting with metals in the processing equipment or container and being left behind, but conclusive information on these points is not available. In dry mixes, it may happen that supplements become unevenly distributed during transfer and filling operations, as a result of differences in density between the supplement and the flour and other mix constitutuents.

VITAMIN FORTIFICATION

In some respects, vitamin supplementation poses fewer processing problems than mineral or protein supplementation because of the smaller quantities of vitamins required. In general, the cost of vitamin supplementation is also less, often just a fraction of a cent per portion, even though the per-pound cost of vitamins is much more than that of most mineral supplements. Cost becomes much more important if Vitamin E and biotin are included at at the high percentages implied by the US RDAs. Processing and storage deterioration of these labile substances must be compensated by addition of an excess so that the consumer receives the full claimed quantity in spite of any decomposition that might reasonably be expected to occur between the time of manufacture and the time of consumption.

There are several sources for most of the individual vitamins, but only a few large firms maintain stocks of all of them. Full-service suppliers will deliver premixes of any specified combination and guarantee their potency. These premixes can be obtained preweighed in batch-size containers or formed into tablets like those often used for bread dough enrichment.

If a product is supplemented with vitamins, or with any nutritional factor, the manufacturer must establish a quality control program to ensure that every portion contains the declared amount of nutrient. In case of an investigation conducted by a regulatory body, it is futile to attempt to excuse a deficiency by saying that an average product (or "nearly all" products) contains the claimed amount; each product must meet the declared content as a minimum.

An excess of certain vitamins can have harmful effects on consumers, and people vary in their susceptibility depending on their age, health, etc. The danger of overdosing is greater with some foods than with others; for example it is highly unlikely that a harmful excess of any vitamin or mineral could be ingested as the result of overindulgence in bakery products, even if the products were supplemented at fairly high levels.

Effects on Color and Flavor

Even though they are used at low levels, vitamins can definitely affect the flavor of foods to which they have been added. Thiamin is the primary offender, but other vitamins can contribute noticeable off-flavors if used in relatively large amounts. Color is generally not a problem, except with riboflavin, which is highly colored. The yellow hue of riboflavin may, however, be compatible with the crumb color of many baked products or be concealed by other colors (as in rye bread or chocolate cake).

Vitamin C, ascorbic acid, can cause a detectable sensation of sourness through its effect in lowering the pH of a food product. If this is undesirable, it may be possible to use vitamin C in combination with its sodium salt to take advantage of the buffering effect that results. Under certain conditions, such as unduly prolonged storage, vitamin C can can react with other ingredients to cause undesirable color changes.

Beta-carotene can change the colors of foods even though it is used in small amounts. As is well known, compounds related to vitamin A, some of them without vitamin activity, are offered and commonly used as pigments to achieve yellowish to reddish hues in a wide range of finished food and beverage products.

Fortification with Folic Acid

There has been heavy pressure on government departments, politicians, and the medical establishment to force folic acid fortification on the public. Leading the charge are various public relations firms, concerned academics, and even some people not paid by the manufacturer of the supplements. The discussion under this heading examines some of the background of this program, the current status of knowledge, and possible future trends.

Neural tube defects — It has been estimated that about 2,500 of the babies born every year in the U.S. suffer from neural tube defects. Neural tube defects include very unfortunate congenital problems such as spina bifida and anencephaly, so there is, of course, an almost universal determination to take whatever means are available to reduce the incidence of such problems. These tragic occurrences result from incomplete closure of the neural tube that has formed in the fetus during the first few weeks after conception.

Many studies have been undertaken to determine the causes and then develop means of preventing these defects. Although unanimity of opinion has not been reached, there seems to be considerable support in the medical community for the view that "folic acid supplementation in the periconceptional period reduces incomplete closure of the neural tube [that normally occurs] at around four weeks' gestation in lean but not obese women" (Goldenberg and Tamura 1996). To clarify this statement, clinical studies indicated that increasing the intake of folic acid reduced the incidence of neural tube defects in infants born to women who were not overweight before and during pregnancy but did not reduce the incidence in women who were obese during this period.

Some questions have arisen, however. There are other influences that may obscure or cancel out the benefits of folic acid supplementation. It was confirmed by Werler et al. (1996) that the risk of neural tube defects in babies increased with increasing prepregnant weight of the mothers, and the incidence of these problems was independent of the effects of folate intake. Further, it was shown by Shaw et al. (1996), the incidence of the defects though related to periconceptual obesity was not related to maternal nonuse of a vitamin preparation containing folic acid, diabetes, use of diet pills, lower dietary folate intake, or a history of NTD-pregnancy. Adjustment for maternal age, education, number of previous pregnancies, use of vitamins, and use of alcohol did not change the odds ratio.

It is reasonable to assume, in view of all these studies, the results of which do not appear to be in dispute, that obesity before, and probably after, conception is a major cause but not the only cause of neural tube defects in the babies.

Mandatory supplementation with folic acid — Reducing obesity in women who are about to conceive would be one approach, but, since obesity is a refractory problem, it appeared to the FDA that a simpler solution would be to add folic acid to everyone's food. In February 1996, the FDA ordered that, by January 1, 1998, enriched bread and rolls, flours, pasta, and certain other grain products be enriched with folic acid to reduce the the risk of birth defects. The rules require the addition by January 1, 1998, of between 0.43 and 1.4 milligrams of the nutrient to each pound of enriched bread, 0.7 mg per lb of enriched flour, enriched macaroni products (including enriched nonfat milk macaroni products and enriched noodle products) 0.9 to 1.2 mg per lb, enriched corn grits and enriched corn meal, 0.7 to 1 mg per lb, enriched farina, 0.7 to 0.87 mg per lb, and enriched rice, 0.7 to 1.4 mg per lb. beginning by January 1, 1998. For breakfast cereals, the addition of up to 0.4 milligram per serving is voluntary, with appropriate labeling.

The new rules allow packagers of foods enriched with folate or folic to make label claims that the nutrient is included and that adequate intake of folic acid has been shown to reduce the risk of neural tube birth defects.

It was noted in the FDA announcement that the consumption of more than one milligram of folic acid per day can mask early symptoms of pernicious anemia, a form of vitamin B_{12} deficiency that can lead to severe nerve damage, including dementia, and thus might cause persons who would otherwise seek help for this dangerous disease to remain ignorant of their at-risk status.

There are some nutritionists who do not think too highly of the prospect of mandatory folic acid supplementation of common foods because of the possibility of serious side effects in certain groups of people. For example, Bostom (1996) said "Given the current state of understanding, I believe it would be highly inappropriate to design secondary prevention trials that assess the long term impact of reducing homocysteine levels on cardiovascular disease outcomes based on folic acid monotherapy ... " Also, Grunwald and Rosner (1996), who state, the " ... suggestion of fortifying grains (flour and cereals) with folic acid to reduce serum homocysteine would also mandate the addition of large amounts of vitamin B_{12} to avoid the risk of precipitating neurologic and/or psychiatric damage to a considerable number of people with undiagnosed and unrecognized pernicious anemia. We believe that it is unreasonable to try to reduce the incidence of coronary artery disease and neural tube defects by measures that could increase the incidence of dementia and neuropathies."

Folic acid is present in many natural foods. As a government publication states (Williams 1994), "A bowl of lentil soup or fortified breakfast cereal, a large spinach salad, or a tall glass of orange juice will put a woman well on her way to 0.4 mg of folic acid." In dark green leafy vegetables, there may be as much as 120 to 160 micrograms of folate in 100 grams of food; fruits (and especially citrus fruits) will often contain 50 to 100 micrograms per 100 grams, and beans and other legumes may contain from 50 to 300 micrograms per 100 grams.

The homocysteine:folate connection — Homocysteine, mentioned in the preceding paragraphs, hitherto has not been a subject of much publicity, even in medical circles. It is regarded as a byproduct and/or intermediate of protein metabolism in humans; it participates in the synthesis of and/or results from the degradation of methionine, an essential amino acid. In addition to being implicated, somehow, in neural tube defects, it has been suggested as a predictor or accompaniment of heart disease and perhaps of other serious disorders. One theory is that homocysteine may make the lining of the arteries more susceptible to damage by increasing the stickiness of blood platelets. Other evidence supports a role for the amino acid in endothelial cell toxicity, effects on coagulation, and stimulation of smooth muscle cell proliferation, all of which might be connected with some aspect of cardiovascular disease.

Morrison et al. (1996) reported the results of a retrospective study to assess the relationship between serum folate levels and coronary heart disease in a panel of over five thousand Canadians. They found a statistically significant association between serum folate level and risk of fatal CHD, generally, the lower the folate level, the higher the risk. They did not find important associations between either use of vitamin and/or mineral supplements *or dietary folic acid consumption* and risk of fatal CHD. They quote other publications that appear to show that folate consumption is a determinant of homocysteine-related carotid artery thickening, the latter being associated with both stroke and CHD.

"Homocysteine and low folate levels are both linked [statistically] with cardiovascular disease risk, yet we do not have proof that increased dietary folate or supplements will reduce the incidence of cardiovascular disease . . . " according to Stampfer (1996), who goes on to say, "Elevated homocysteine can be reduced by even modest amounts of folate, providing a plausible mechanism for the remarkable findings of Morrison et al. of a 69% increased risk of coronary mortality among those in the lowest quartile, as compared with the highest quartile of serum folate."

The relationship of serum levels of homocysteine to risk of cardiovascular diseases may be suggestive, but the connection to any dietary factor has not been clearly established. Serum folate levels are not closely associated with dietary folate. On the other hand, the excessive amount of homocysteine in the blood of certain heart patients apparently results from gene mutations causing enzymes to be produced that trigger synthesis of the compound, according to a publication in *Nature*, quoted by Katz (1996). Once again, we see the overriding factor of genetic constitution on susceptibility to diseases.

Safety of Micronutrients

It is well known that most of the micronutrients can be toxic if ingested in amounts far exceeding the prescribed dose, but it is probably not as widely recognized that some of the trace minerals (particularly) can have adverse effects at relatively low multiples of the recommended intake. The problems of iron enrichment have been mentioned previously; in the following section certain other elements will be considered (discussion based partly on Katz 1996).

The essential nutrients regarded as trace minerals are copper, chromium, cobalt, fluorine, iodine, iron, manganese, molybdenum, nickel, selenium, vanadium, and zinc. At a zero dose of an essential element, adverse health effects are observed. As the does increases, the adverse effects are ameliorated. At an optimum level, different for each element, no adverse effects are noted. Following this is a tolerance zone, which no toxicity symptoms can be observed. At higher doses, toxic signs are noted. Of course, all ingestibles have some toxicity; the LD_{50} for salt is 4 g/kg for mice. For elements regarded as toxic by nature, e.g., arsenic, cadmium, chromium, lead, mercury, nickel, and selenium, the toxic effect begins to be observed at a very low level. The reader will note that chromium, nickel, and selenium are listed as both essential and toxic, indicating that any supplementation with these ingredients should be carefully judged in the light of possible harm to persons who may have special sensitivities to these materials or who are already ingesting (under whatever unusual set of circumstances) an ample amount of the mineral.

The RDA for zinc is 15 mg per day and the LD_{50} is 2 g/kg for mice. The chance that such a level could be approached through the intake of any supplement expected to be made available is highly unlikely, but the possibility that adverse health effects could occur at much lower levels must be taken into consideration, especially for persons already in poor condition. Reproductivity efficiency is thought to be decreased by zinc deficiencies, and medication or supplements known to affect this group of physiological actions have a history of being heavily overconsumed, especially by males.

The NRC has not established an RDA for copper, the estimated safe and adequate daily dietary intake is 1.5 to 3 mg. Many soluble copper compounds have an LD_{50} of about 50 mg/kg for mice. This is a fairly low level, and indicates great caution should be used in attempting to, or recommending, copper supplementation beyond that normally present in a food or ingredient because possible ill-effects on consumers who are particulary susceptible to the element or who, in the course of their normal activities (drinking large amounts of water high in copper, for example) take in much more than the expected amount of water.

Selenium has been classified both as an essential trace element and as a toxic element. The RDA for males is 55 µg/day for females and 70 µg/day for males. The Environmental Protection Agency has established a maximum contaminant level for selenium in potable water of 10 µg/L; for an average daily consumption of two liters of water, a dose of 20 µg/day would be reached. The LD_{50} for rats is 3 mg/kg. Certainly, there is not much room for error in supplementing with this element.

According to Katz (previously cited), "The essentiality-toxicity aspects of chromium are equally confounding. The estimated safe and adequate daily dietary intake [ESADDI] for an average 70 kg adult human is from 50 to 200 μg, which corresponds to an approximate daily dose of 2 μg/kg. The chromium potable water maximum contaminant level established by EPA is 100 μg/L; the consumption of two liters per day is equivalent to a daily dose of 200 μg. On the basis of the maximum contaminant level and the ESADDI, it appears that the 200μg/day dose corresponds to both the upper and the lower limit for chromium intake." The valency of the chromium atom makes a large difference in toxicity of the element; hexavalent chromium compounds appear to be at least 20 times more toxic to rats when administered orally. Chromium is also a carcinogen. Obviously, much thought should be given to the possible health disadvantages as well as advantages when considering the addition of a chromium source to a supplement for man or beast.

Stability of Micronutrients

Some of the vitamins are very reactive compounds, and are apt to deteriorate through interactions with other ingredients, oxygen from the air, packaging materials, and other agents in their immediate environment. Proteins, peptides, and amino acids may also undergo undesirable changes, particularly through the so-called Maillard, or nonenzymic browning, reactions with reducing sugars. Mineral additives are less susceptible to chemical changes, although in a few cases change of state or ionic status can reduce physiological availability of the desired element.

Effects of type of food — The stability of vitamins and some mineral additives is influenced by the physical and chemical properties of the other ingredients individually as well as by the bulk environment in which the additive is placed. Components such as nitrogenous compounds, fibrous materials, and reducing sugars could affect stability and bioavailability of nutrients. The forms of iron that are most bioavailable generally are the most reactive and may lead to the development of organoleptic defects in finished products. Canned foods offer opportunities for the contents to react with the liner and sealants, allowing decomposition of nutrients and, in a few cases, deterioration of the liner to such an extent that corrosion of the can body may occur.

Effects of packaging — Packaging design should take into consideration the inherent sensitivities of the added nutrients. Vitamin C and carotene should be protected from oxygen. In a beverage, oxygen can quickly degrade vitamin C, so hermetic seals and packaging materials impervious to oxygen transfer are highly desirable. Radiation, including sunlight and fluorescent illumination, is generally damaging to vitamins, suggesting the advantages of using opaque containers.

Effects of processing conditions — Time and temperature of the heat processing procedures used to cook, pasteurize, or sterilize should be kept to the minimums necessary to obtain the essential results. Most minerals are not adversely affected by heat but many vitamins are unstable at high temperatures. Short-time high-temperature methods are generally less destructive to vitamins than are long-time medium-temperature procedures. The unit procedure of freezing does not generally cause deterioration of these nutrients, but the preliminary steps of washing and blanching can cause losses of water-soluble vitamins and minerals.

A substantial percentage of some of the vitamins may be destroyed because of the conditions they encounter during baking. Vitamin C is particularly labile, and vitamin B_1 is fairly sensitive, while niacin is relatively stable.

To prevent undue loss during processing, the vitamin supplement can be sprayed on some baked products after they have cooled — possibly this method would be suitable only for nonstandard products. If this application method is selected, concentrated aqueous suspensions should be used to avoid the adverse effects on texture, appearance, and stability that would result from contact of the product with excess water. Consideration should be given to spraying the bottom of loaves, rolls, cakes, etc. to avoid visible defects. An oil suspension might be used. Dusting with a dry mixture of the vitamins might leave a detectable residue and presents difficulties in securing uniform distribution. If the product has a filling or icing, the enrichment might be added to it to avoid the more extreme condtions (such as oven temperatures) encountered when the additive is in doughs and batters.

Vitamins can be obtained in encapsulated form; this reduces the odor and flavor problem as long as the additive remains dry and it reduces storage deterioration to some extent. Encapsulation is of little value if the vitamins are put through a wet-processing step, especially if the blend is also heated. Fat-soluble vitamins are often supplied as oil solutions or in minute pellets composed of fatty substances, in which case it is advisable to specify an antioxidant content since the corn oil vehicle sometimes used can become rancid eventually.

Shelf life — Products expected to spend a long time in the distribution system and on the store shelf will require higher overages of some nutrients such as vitamins A and C so as to meet label claims up to the time of their pull dates. The conscientious manufacturer will want to use additions that are large enough to give the expected and declared dosage at the time of the product is consumed, even allowing for some loss between purchase and consumption, i.e., while the product is in the user's possession. The average consumer pays little attention to expiry dates once he has purchased a food, with the possible exception of dairy products. Most minerals exhibit good storage stability, and a few vitamins (such as vitamin E and pyridoxine) also pose few problems under normal conditions.

Storage conditions — Time, temperature, and humidity are the major environmental factors affecting storage deterioration of added nutrients. With canned foods, temperature of storage is the overriding factor. Frozen foods can be expected to retain nutrients better than similar products stored at higher temperatures but deteriorative reactions occur even in these items. If foods in transparent packages are exposed to fluorescent lighting or sunlight, loss of nutrients may occur at relatively high rates.

FORTIFICATION WITH PROTEINS OR AMINO ACIDS

For persons in the United States subsisting on diets falling within the normal range of variation, protein availability does not seem to be a particular problem. However, when consumers deliberately restrict their consumption of calories to a very low level, or follow some highly selective dietary regimen, such as strict vegetarianism, the intake of essential amino acids may fall below the level necessary for maintaining body functions. Persons trying to lose weight at a rapid rate by cutting their caloric intake to very low levels may also need to consider using protein supplements. The commercial value today of claims of protein supplementation is doubtful, but in the not too distant past there was a considerable amount of publicity about the need to add protein to the diet, and products were developed to fill this apparent need. These fads seem to go in cycles, so the future may bring us back to a recognition of the desirability of adding protein to normal diets.

In some other nations of the world, and among certain groups in this country, the lack of adequate intake of essential amino acids continues to be a problem. Protein malnutrition may affect people who must maintain themselves on food that consists almost entirely of cereals or other starchy foods, or who have difficulty in obtaining satisfactory amounts of food of any type. Those persons who are strict vegetarians by choice may also have difficulty in consuming enough protein to satisfy their basic needs.

Because bread-like products are staple foods in many countries, it may be desirable to increase protein intake by adding protein fortifiers to wheat flour or to the dough. Alternatively, specific amino acids in purified form can be added to improve the PER of the cereal proteins. Lysine is the limiting amino acid in many cereal products. So far as its marketing advantage is concerned, however, protein supplementation no longer seems to be in the top rank of consumer interests in the U.S. Other types of nutritional manipulation have taken the consumer's fancy.

It should not be very difficult to add a few percent of protein to pasta, if those products are the preferred bulk foods of the area in question. Some visual clues to the enrichment's presence would be likely, but careful selection of the ingredient should keep flavor and texture changes to a minimum. Rice and potatoes pose more difficult problems. Imitation rice made in the same general way that pasta is made, i.e., extrusion forming, is a possibility. The gruels or porridges made from taro, sorghum meal, and the like, are relatively structureless foods that vary in flavor from batch to batch and time to time, so their supplementation should be relatively simple.

Table 9.3
RECOMMENDED DIETARY ALLOWANCES FOR PROTEIN

Category	Age (yr)	Weight (kg)	RDA (g/kg)	RDA (total)
Both sexes				
	<0.5	6	2.2	13
	0.5-1	9	1.6	14
	1-3	13	1.2	16
	4-6	20	1.1	24
	7-10	28	1.0	28
Males				
	11-14	45	1.0	45
	15-18	66	0.9	59
	19-24	72	0.8	58
	25-50	79	0.8	63
	51+	77	0.8	63
Females				
	11-14	46	1.0	46
	15-18	55	0.8	44
	19-24	58	0.8	46
	25-50	63	0.8	50
	51+	65	0.8	50

Source Federal publications.

Nutritional Requirements for Proteins and Amino Acids

There are two major points to be considered when discussing the adequacy of protein nutrition. One is the total amount of protein that is available to the consumer, the other is the amino acid distribution afforded by the proteins. Humans need a certain amount of protein to supply the store of amino acids that can be utilized as is or converted into other amino acids for which there is a greater need. In addition, there are some amino acids that cannot be synthesized by the human system, and they must be obtained from the food ingested by the person or a healthy status cannot be maintained. Many scientific studies have been conducted to determine the total amount of protein needed for maintaining good physiological status in humans. The available data has been reworked by scientists to provide the data in Table 9.3, showing the recommended minimum intake of protein for persons of different age groups.

Not given in the Table 9.3 are the added requirements for pregnant or lactating women. To the basic allowance, there should be added ten grams per day for the duration of pregnancy. For lactating women, fifteen grams of protein should be added per day during the first six months and twelve grams per day during the second six months. Other factors can also modify the need for protein — illness, heavy work or intense exercise, and individual idiosyncrasies.

Table 9.4
COMMON NATURAL AMINO ACIDS

Essential amino acids	Non-essential amino acids
Histidine	Alanine
Isoleucine	Arginine
Leucine	Aspartic acid
Lysine	Asparagine
Methionine	Cysteine
Phenylalanine	Cystine
Threonine	Glutamic acid
Tryptophan	Glutamine
Valine	Glycine
	Hydroxyproline
	Proline
	Serine
	Tyrosine

Knowing the amount of protein needed does not tell us all we need to know about nutritional requirements. Researchers have done much work to determine which of the amino acids cannot by synthesized by normal individuals and therefore must be obtained from food. Table 9.4 lists the amino acids that are currently believed to be "essential," that is, they cannot be synthesized by humans, at least not in the amounts needed for proper functioning of the body. There are currently thought to be nine essential amino acids.

There are about thirteen other common amino acids, not regarded as "essential," and other compounds of this general sort are known to exist in proteins found in nature, but they are not usually taken into account in studies on human nutrition. Of course, amino acids are normally ingested in the form of proteins, and must be converted by digestive enzymes and other enzymes into short chain peptides and amino acids for further conversion and modification by body processes.

If a person has serious damage to body processes (e.g., liver disease) or genetic insufficiencies or for some other reason (premature infants) cannot process amino acids in the normal pattern, certain of the otherwise non-essential amino acids may be required, i.e., they become essential to survival. Among these are cystine, tyrosine, and arginine.

Table 9.5
AMINO ACID REQUIREMENTS FOR DIFFERENT GROUPS[1]

Amino acid	Infants 3-4 mos.	Children <2 yrs	Children 10-12 yr	Adults
Histidine	28	--	--	8-12
Isoleucine	70	31	28	10
Leucine	161	73	42	14
Lysine	103	64	44	12
Methionine + cystine	58	27	22	13
Phenylalanine + tyrosine	125	69	22	14
Threonine	87	37	28	7
Tryptophan	17	12.5	3.3	3.5
Valine	93	38	25	10

[1] In mg of amino acid per kg of body weight, per day.
Source: Sakai and Kawakita (1991) and Smith (1992). These are estimated average requirements for healthy subjects; they are not federal requirements.

Human requirements for average daily intake of the individual essential amino acids have been reasonably well established by many studies over a period of many years. Table 9.5 gives the present consensus on these requirements for different age groups. These amounts are not contained in any federal regulations covering foods. In fact, there does not seem to be any federal policy on the need for, or labeling information for, individual essential amino acids.

Like many other organic compounds, amino acids occur as L- and D-isomers. Almost all natural amino acids occur as the L form, although rare occurrences of the D isomers have been reported. The body's physiological processes are based on the L isomer and so this type of amino acid is more readily absorbed by the body and provides more nutritional value than the other configuration. When amino acids are made synthetically, the product is usually a mixture of the two types, i.e., a racemic mixture.

The term PER is prominent in the scientific literature concerned with the effectiveness of individual proteins in human nutrition. This term (an abbreviation of Protein Equivalency Ratio) indicates data obtained in an animal assay of the efficiency of a protein (or amino acids) in promoting growth. Usually, a test using rats as the subjects is conducted. In these tests, the protein efficiency of a sample is compared with that of the reference material, casein, which is taken to have a PER of 2.5. The PER is then computed by dividing the weight gained by the amount of protein consumed by the animal. It is noteworthy that a food with a PER of, say 2.0, is not necessarily twice as good as a protein having a PER of 1.0.

There are many methodological and conceptual problems with the PER test, though it has been useful in providing insight into protein differences during the many years it has been accepted. Scientists are now studying the possibility of combining the amino acid score (based on a determination of the amino acid composition of the food) and a determination of the digestibility of the protein.

Natural proteins are absorbed by the digestive system and metabolically utilized at different degrees of efficiency. These differences must be taken into account when estimating the value of the protein contained in a food to the food's importance in diet. Digestibility can be determined using either *in vitro* or *in vivo* methods.

First, the milligrams of essential amino acids in one gram of test protein are determined, then the results are divided by the amount of the compound found in a reference standard. Finally, the amino acid score is derived from the value of the limiting amino acid, this being the amino acid exhibiting the lowest concentration. Protein values as a percent of the RDI that may be listed on the labels of products covered by the NLEA must be determined using the corrected amino acid score method.

Table 9.6
AMINO ACID SCORES FOR GRAINS AND OTHER FOOD MATERIALS[1]
CORRECTED FOR PROTEIN DIGESTIBILITY

Food or ingredient	Protein %	True P%D	Amino acid score	P%D corrected score
Casein	94.7	99	1.19	1.00
Egg white	87.0	100	1.19	1.00
Pea flour	30.8	88	0.79	0.69
Peanut meal	61.2	94	0.55	0.52
Rapeseed protein isolate	87.3	95	0.87	0.83
Soy protein concentrate	70.2	95	1.04	0.99
Soy protein isolate	92.2	98	0.94	0.92
Sunflower protein isolate	92.7	94	0.39	0.37
Wheat gluten	87.	96	0.26	0.25
Whole wheat	16.2	91	0.44	0.40
Rolled oats	18.4	91	0.63	0.57

Sources: World Health Organization, and R. Smith 1992.
[1] Dry weight basis. P%D=Protein percent digestibility. Protein=N x 6.25.

The World Health Organization has recommended that a combination of amino acid scores and *in vivo* digestibility values be used as the best measure of the overall quality of a protein food. See Table 9.6 for examples of amino acid scores for some common food ingredients, and the corrections applied for digestibility.

Justification for Supplementation

Promotion of fabricated foods containing added protein should not be undertaken unless marketing tests show the targeted population recognizes a need for this kind of product. Although large-scale agitation and publicity campaigns urging donations of massive amounts of proteinaceous foods to underfed foreign populations has decreased markedly in the 90s, there are still opportunities for distributing protein-enhanced foods in some government programs here and abroad.

Supplementation of fabricated foods, such as bakery products, with significant amounts of protein is difficult because most of the available additives alter one or more of the organoleptic aspects of the finished item.

Protein supplementation is also expensive; unlike vitamins which, though costly on a per pound basis, are added in micro-amounts, several grams of protein must be added to each portion to achieve an increase sufficient to be claimed on the label. Many supplements which would be excellent sources of protein have found limited acceptance because of high cost, functional deficiencies, or poor organoleptic properties.

Protein supplementation takes two forms: (1) Adding purified amino acids or complementary proteins to increase the PER (an indication of the nutritional quality) of the proteins normally present in the food, and (2) increasing the total amount of protein present by adding some kind of concentrated nitrogenous material such as casein, whey protein concentrates, isolated soy protein, or egg white. For labeling purposes, the second alternative is preferred.

Protein supplements must be added to staple foods at the parts per hundred level in order to make a substantial improvement in the nutrient intake, as compared to parts per million for some of the vitamins and minerals. Thus, cost becomes a very important factor. Effects of the supplements on the physical properties of processing intermediates (such as doughs and batters) and on the organoleptic properties of the finished products are often limiting factors.

Although there are many kinds of concentrated proteins that have been suggested for use in enriching bakery products, in most cases economic factors and organoleptic considerations tend to narrow the options to soybean and milk derivatives. Legal restrictions and labeling problems also operate to narrow the baker's choices. Some of the protein-enhancing ingredients available to the food product formulator will be discussed in the following paragraphs.

The ideal protein supplement for adding to fabricated foods such as bakery products would be a cheap, bland, white powder with a high protein efficiency ratio (or, at least, should contain relatively large percentages of the essential amino acids that are present in low concentrations in the other ingredients). It should not have unpleasant ethical or esthetic connotations. Since there does not seem to be any natural animal or vegetable derivative that fulfills all of these requirements, compromises must be made. Dried egg white probably comes closest to the ideal, but it is very expensive.

Soy protein preparations are used far more than any other protein supplement, mainly because of their relatively low price and their ready availability. Next in importance are the milk protein preparations made from whey and casein. Egg proteins are generally too expensive for this purpose unless they also contribute other important properties to the finished food.

Vegetable-derived Protein Supplements

In addition to their economic advantages and almost unlimited supply potential, protein supplements derived from seeds and other plant sources are especially appealing to food formulators because they are acceptable to vegetarians and avoid the meat/dairy prohibition in kosher cuisine. Cost is also less (usually) than additives derived from animal products and chemically synthesized materials such as lysine and methionine. Although many seeds and seed derivatives have been studied to determine their suitability for use as protein supplements, preparations from soybeans dominate this market.

Soy proteins — Defatted soy flour and grits are among the most economical sources of soy protein available. The nutritional value of these ingredients is somewhat superior to that of the concentrated (refined) protein products made from soybeans. That is to say, the protein of the flour and grits has a slightly higher PER than the more highly purified fractions. Soy protein isolates and concentrates have the considerable advantage of causing less damage to flavor, color, and texture than the defatted flour and grits. In general, however, the organoleptic properties of most foods decline in direct proportion to the amount of these enrichments added to the dough.

The production of soybean-protein ingredients for foods generally starts with the oilcake that remains after oil has been extracted from the seeds. Whole fat meal is seldom used because of its higher cost and the limited storage stability that results from the oil's susceptibility to oxidative rancidity and flavor reversion. Soybean flakes (ground cake) normally analyze as more than 50% protein, and this protein contains enough lysine to make it a good nutritional complement to the proteins in wheat flour. Thus, the mixture of wheat flour and defatted soybean meal has a greater nutritional value for humans than would be predicted from a simple summation of its parts.

Defatted soy flour can be made simply by grinding the oilcake. The raw material is usually heat-treated before the flakes are ground. Flour that has not been treated has considerable enzyme activity; in fact, it has been used in bakery formulas at up to 0.5% of the flour weight, with the major purpose of lightening the color of the finished product and improving dough properties through operation of the lipoxidase that is present in it. At these addition levels, the protein does not a significant influence on the nutritional properties of the product, however. Larger percentages almost always cause problems with flavor.

Soy concentrates containing about 70% protein are made by extracting most of the unwanted soluble materials from defatted soybean flakes. In a typical process, the flakes will be extracted with aqueous alcohol, dilute acid, and hot water. The substances that are removed are mostly unfermentable sugars and other higher saccharides. The insoluble materials that remain are neutralized, dried, and ground to make the ingredients offered for food supplementation. Pound for pound, the protein in soy concentrates is considerably more expensive than protein in the form of heat-treated, defatted soy flour.

The manufacture of soy isolates, which contain about 90% protein, starts with defatted meal. Protein is solubilized with an alkaline aqueous solution, the liquid is separated, and acid is added to precipitate most of the protein in the form of a curd. This precipitate is separated, washed, neutralized, and dried to give a powder. Soy isolates are relatively bland in flavor and light in color, and are rather easily dispersible in aqueous systems if clumping can be prevented. They are also relatively expensive.

The soy product that finds the greatest employment as a protein enhancer is the soy concentrate. Soy isolates can be used, and are in many respects better than the concentrates, but they are usually too expensive to be considered as a practical alternative.

Heat treatment improves the flavor, removes antinutritional factors, and inactivates enzymes in soy products. It also reduces solubility of the protein, and the extent of this effect has been expressed as the "Protein Dispersion Index" or PDI. Unheated flour will have a PDI of more than 90; moderate heating will reduce this to about 70, and heat treatment sufficient to toast the flakes will give a PDI of around 20. Meals of 70 PDI are widely used at levels of up to about 10% (FWB) as protein supplements. The 20 DPI meals are dark in color and somewhat "nutty" in flavor, and are used mostly for color and flavor enhancement in crackers and variety breads.

Soy flours with PDIs of around 70 are often used in combination with dried whey and minor amounts of other ingredients as milk replacers in white breads. These mixtures give some of the same dough and finished product characteristics as nonfat dry milk at a lower cost. Strictly speaking, this is not a protein supplementation use. If milk is removed and "milk replacer" added, the net effect on protein content of the bread or other baked product will depend largely on the protein content of the replacer. The ingredient's protein content varies between brands and types but seldom results in a significant increase in net protein content of the finished product when the substitute is used to replace nonfat dry milk (on an equal weight basis) in a product formulation.

The information in Table 9.7 summarizes the composition of four kinds of supplements derived from soybeans: defatted soy grits, textured vegetable protein, soy protein concentrate, and soy protein isolate. Materials of this sort are readily available from several commercial sources. Flavor, odor, texture, and appearance will vary between suppliers and between different brands from the same supplier.

Table 9.7
PROTEIN SUPPLEMENTS FROM SOYBEANS[1]

Characteristic	Defatted soy grits	TVP[2]	Soy protein concentrate	Isolated soy protein
Proximate				
Moisture, %	9	6	6.0	6.0
Protein, (Nx6.25), %	52	52	70.0	91.5
Fat, %	1	1	0.5	0.5
Ash, %	6	6	5.3	4.5
Crude fiber, %	3	3	3.5	0.5
Carbohydrates, %	29	32	17.7	--
Other Analytical				
PDI	20	--	--	--
Trypsin inhibitor, mg/g	3	--	--	--
P.E.R.	2.2	2.2	--	--
Calories per 100 g	280	280	352	380
pH	--	--	--	7
Essential Amino Acids[3]				
Lysine	6.2	6.1	6.3	6.1
Threonine	4.1	4.2	4.0	4.0
Leucine	7.7	7.8	7.7	8.2
Isoleucine	4.6	4.7	4.5	4.9
Valine	4.7	4.8	4.8	5.2
Tryptophan	1.2	1.1	1.5	1.1
Phenylalanine	5.0	5.0	5.3	5.0
Tyrosine	3.5	3.3	3.6	3.6
Methionine	1.2	1.2	5.3	1.4
Cystine	1.5	1.5	1.4	1.4
Histidine	2.7	2.7	2.6	2.7

[1] Notes: Fat is the petroleum ether extract. Carbohydrates, by difference. PDI = Protein Dispersibility Index by AOCS method Ba 10-65. PER = Protein Equivalency Ratio, casein being 2.5. Isolated soy protein is Ardex D brand. Soy protein concentrate is Arcon F brand.

[2] TVP = Textured vegetable protein, unfortified. ADM brand. Fiber by the Neutral Detergent Method is 22%

[3] In grams of amino acid per 100 grams of protein.

Source: Various commercial brochures.

Proteins from other vegetable sources — There are several plants other than the soybean that are grown primarily for their oil content. When oil has been pressed or solvent-extracted from the seeds, there remains a residue that contains other materials such as carbohydrates, fiber, and protein.

The so-called oilcakes are mostly used as protein sources in mixed animal feeds, but sometimes they are suitable for use only as fuel or fertilizer. Cottonseed, a byproduct of a plant grown primarily for its fiber, is also an important commercial source of oil and cake.

Some oil seeds of worldwide commercial significance are sunflower, rapeseed, peanut, sesame, flax, and cottonseed. There have been many attempts to process the oilseed cakes from these materials into food ingredients, with variable success. Protein isolates from cottonseed and peanuts have been used in foods on a commercial scale.

Using high protein wheat in the mill mix has sometimes been suggested as the ideal way to increase protein content in bread, but it does not seem to be a practical method because the suitability of flour for baking purposes can be changed drastically by a few percentage points difference in protein content of the flour. Wheat mixes blended for the milling of bread flour tend to have protein contents in a narrow range and a typical and rather constant amino acid distribution. The pattern of amino acids depends more on the total nitrogen of the sample than on the particular variety of wheat. When calculated on an equal nitrogen basis, samples of wheat with high or low nitrogen contents differ but slightly in the relative amounts of certain amino acids.

The residues remaining from production of flour, semolina, farina, rye flour and meal, corn meal, and other granular or powdered cereal products, are collectively called millfeeds. They have fairly high protein contents and are mostly used as components of cattle rations. These materials do have some value as supplements for specialty food products, but their high percentage of fiber (which may be an advantage in some applications), their pronounced and often unpleasant flavor, their limited storage stability, their dark color, and their physical properties create obstacles to such usage. The same objections in accentuated form apply to the residues from brewing and from the corn wet-milling process. It might be possible to extract bland nitrogenous substances from these byproducts, with or without prior enzyme digestion, but only a limited amount of work seems to have been done along this line.

Protein Enhancers from Animal Sources

The supply of protein enhancers from animal sources is potentially very large, from a theoretical standpoint, but the actual usage is essentially restricted to ingredients obtained from milk or eggs. Attempts have been made to introduce protein from blood serum obtained from slaughterhouse operations, etc., but these have not been successful, largely (it is believed) from esthetic considerations. Concentrates made from meat scraps have been suggested, but cost and non-uniformity have prevented their acceptance. Gelatin should be mentioned; it is of course widely used as a component of jellied desserts, confections, etc. It is a protein of sorts made from pork skins, cattle bones, and the like, but its nutritional value is very low since it has a high content of glycine and very low content of essential amino acids. It was at one time recommended for various nutritional uses, such as strengthening fingernails, but is no longer regarded as very useful nutritionally. It has important functional uses in confectionery, desserts, etc., where its gelling capability, bland flavor, and light color endear it to formulators. It also has many industrial applications.

Milk proteins — Whole milk in any form is too expensive to be used primarily as a protein enriching agent. In addition, its fat content boosts the calorie count to a high level. Nonfat dry milk is a fairly good source of protein, but it, too, is rather expensive and it carries a large percentage of carbohydrate (i.e., lactose) that makes it difficult to fit into some formulations. Dried whey contains an even higher percentage of carbohydrate than skim milk powder, but it is relatively inexpensive and it avoids some of the functional disadvantages of nonfat dry milk, especially when the whey has been properly heat-treated.

Concentrated protein supplements obtained from milk are widely used. They are generally categorized as either serum (whey) proteins or casein. Both these materials have high PERs making them more effective dietary supplements, pound for pound, than soybean protein. They are also more expensive than vegetable protein concentrates, as a rule. Although whey powder itself is of little value for protein supplementation, since it contains only 12% to 13% protein combined with high levels of lactose and minerals, methods have been developed for separating the protein. The resultant protein concentrate is bland, if it has been prepared from raw materials of good quality, but it may have unpleasant odors and flavors when prepared from whey that has undergone bacterial action or other forms of deteriorations.

Whey protein concntrates have been prepared commercially by electrodialysis, ultrafiltration, Sephadex gel filtration, and complexing with metaphosphates. Protein concentrates of the dried materials range from about 26% to about 58%, and can generally be adjusted by altering the fractionation conditions. The higher the protein content of a concentrate, the higher per-unit cost of protein it contains, generally speaking.

The proteins in concentrates prepared from whey will have undergone variable amounts of denaturation. The extent of denaturation is related to the heat treatment the proteins encounter during processing and it has an effect on the ingredients' solubility and other functional properties. Whey intended for use in bakery products is deliberately given a high-heat treatment to reduce the ingredient's softening (weakening) effect on gluten. Denaturation of this degree has little or no negative effect on the nutritional properties of the protein.

Whey protein has an excellent PER. Commercial whey concentrates will interfere only slightly with most of the handling properties of doughs into which they have been incorporated if the whey has been properly heat-treated, and they are relatively inexpensive. Unfortunately, some of these ingredients have undesirable tastes or odors that can carry through to the finished product. Careful screening of suppliers and rigid policing of deliveries is absolutely essential when buying whey powder or any whey protein concentrate.

Casein for protein supplementation is obtained from skim milk by acid precipitation, a reaction similar to that used in preparing certain kinds cheese curd. The curd is washed, dried, and ground to give the commercial product. It is poorly soluble, but can be used in applications where this property is not a problem. Casein is used for protein supplementation to an extent varying with the economics of the milk market. It can cause some flavor problems because of the tendency of manufacturers to use as their raw material milk that has become unsuitable for consumption as a beverage. The PER of casein is very good. Sodium caseinate is easier to disperse in liquid than is casein, but it is far from ideal in this respect.

Egg products – Egg products not only contain high percentages of protein, but the proteins have very high PERs, making them excellent sources of this nutrient. Egg derivatives have the potential of being the most nutritionally effective of all readily available concentrated proteins. Unfortunately, eggs have gained the image in consumer's minds of containing a lot of cholesterol, a substance that many people believe is connected with health problems. Egg yolk is indeed fairly high in cholesterol, as compared to, say, skim milk, but egg white contains essentially no cholesterol. For that matter, egg white contains no significant amount of any kind of lipid.

Although fresh eggs are bland in flavor, and egg white markedly so, they tend to deteriorate fairly rapidly in both flavor and physical properties even when stored at refrigerator temperatures. Dried whole eggs are reasonably stable, and dried egg white very stable at room temperature, provided the small amount of glucose ordinarily present has been removed by enzyme treatment and, of course, provided the moisture content is kept sufficiently low. Most dried egg derivatives have been enzymically treated to inactivate glucose.

Eggs also contain useful amounts of several vitamins, but the mineral content is not outstanding. The division of vitamins between the white and yolk is interesting, the yolk containing relatively large amounts of thiamin, pantothenic acid, vitamin B_6, vitamin B_{12}, and vitamin A, while the albumen contains much smaller amounts of these components, but significantly larger amounts of riboflavin and niacin,

Table 9.8 summarizes the content of eight common vitamins and protein in fresh eggs and the separated yolks and albumen, while Table 9.9 contains the same information for dried eggs. Notice the remarkably high content of protein in dehydrated egg white, over 82% on an as is basis, and about 90% on a bone-dry basis. Comparable figures for dried whole eggs are 45.8% and 47.8%. It is clear these materials could be very effective agents for increasing the protein content of other foods.

Table 9.8
PROTEIN AND VITAMINS IN FRESH EGGS[1]

Constituent	Whole eggs	Egg white	Egg yolk[2]
Vitamins			
Thiamin, mg	.087	.005	.254
Riboflavin, mg	.301	.285	.436
Niacin, mg	.062	.089	.069
Pantothenic acid, mg	1.727	.241	4.429
Vitamin B_6	.120	.003	.310
Folacin, mg	.065	.016	.152
Vitamin B_{12}, mcg	1.547	.065	3.803
Vitamin A, IU	520	0	1,839
Other Constituents			
Protein (Nx6.25), gm	12.14	10.14	16.40
Water, gm	74.57	88.07	48.76

Source: U.S. Agriculture Handbook 8 and elsewhere.

[1] Amounts expressed in milligrams, micrograms, grams, or International Units (as indicated) per 100 grams edible portion, as is. Products are generally spray-dried commercial products without additives, but the white and whole egg are glucose-reduced by enzyme treatment.

[2] Fresh yolk contains a small proportion of white.

Dehydration of eggs by standard commercial processes does very little damage to the PER and results in very little change in the protein content (when the latter is calculated on a dry matter basis). Some commercial dried egg products contain additives intended to improve the product's dispersibility, stability, whipping properties, etc., and these diluents must be taken into account when calculating the nutrient content of ingredients. Although the chemical effects of such additives has never been studied (or at least has not been reported), it is not expected that they would have much destructive effect on either protein or vitamins. In fact, if they prevent or reduce non-enzymic browning reactions, the PER after storage should be improved, i.e., should be higher for treated than for non-treated products.

Table 9.9
PROTEIN AND VITAMINS IN DEHYDRATED EGGS[1]

Constituent	Dried whole eggs	Dried egg white	Dried egg yolk[2]
Vitamins			
Thiamin, mg	.325	.037	.435
Riboflavin, mg	1.232	2.316	.811
Niacin, mg	.259	.723	.128
Pantothenic acid, mg	6.710	1.958	8.242
Vitamin B_6, mg	.420	.024	.577
Folacin, mg	.193	.096	.213
Vitamin B_{12}, mcg	10.51	.528	7.08
Vitamin A, IU x 10^{-3}	2.05	0	1.84
Other Constituents			
Protein (Nx6.25), gm	45.83	82.40	30.52
Water, gm	4.14	8.54	4.65

Source: U.S. Agriculture Handbook 8 and elsewhere.
[1] Amounts expressed in milligrams, micrograms, grams, or International Units (as indicated) per 100 grams edible portion, as is. Products are spray-dried commercial products without additives, but the white and whole egg are glucose-reduced by enzyme treatment.
[2] Yolk contains a small proportion of white due to commercial separation procedures.

Finished products of high protein value — Meringues may have some value as nonfat snacks or as a relatively low calorie sweet snack. They have the advantage of a sweet taste without other obtrusive or cloying flavors, and easy and rapid digestibility.A typical meringue formula and procedure follows. Such materials are very satisfactory for topping cream pies, but are a little too weak structurally and not dense enough for confectionery bases. A meringue of slightly lower moisture content, formed into cookie shapes and baked until almost dry, will have good stability and be sufficiently firm to allow packaging and shipping through normal channels. The lower moisture content will also increase the "as is" protein percentage, of course.

Formula

Spray dried egg white, containing whipping improver, 2.8%
Tapioca starch, 2.6%
Powdered sugar, 93.3%
Pectin, 0.4%
Gelatin, 0.3%
Cream of tartar, 0.3%
Citric acid, 0.1%
Salt, 0.2%
Water, about 62 parts to 100 parts of the above mixture.

Function of the Ingredients

The egg white and water form the basic foam structure, when properly whipped. The pectin and gelatin assist in forming the cold structure and maintaining it through application and baking. The tapioca starch also adds greater rigidity to the foam, especially after heating. The cream of tartar and citric acid form a buffering system that provides a pH in the water phase that is ideal for extending the egg protein over large surfaces, i.e., improving the egg white's whipping response.

The sugar and salt are taste improvers, and the former is an important bulking agent. A top-note flavor, preferably a good grade of concentrated vanilla, should be added to the mixture at a level meeting the needs of the intended application; very little if the meringue is to be used as a topping on cream pies.

Procedure

Prepare a blend of the dry ingredients. Add water in the proportion of one pound water to 1.6 dry blend. Mix slowly, then beat with a whip at high speed until a stiff peak is formed.

Results

This confection cannot be regarded as a good source of protein, since it contains only about 1.5 or 1.6 percent protein in the wet product (the gelatin should not be regarded as nutritional protein), but the protein it does contain is of very high quality. The protein content can, however, be boosted without significantly harming the organoleptic properties of the finished product. Meringues are devoid of fat and cholesterol, and are relatively low in calories compared to, say, whipped cream. It is possible to make a salt-free version, if desired. Even in the unmodified form, the sodium content is not high.

Angel food cakes are better sources of protein than are meringues, but even so, they are only moderate in this respect. The following commercial formula is typical of the two-stage method.

Formula — first part
Spray dried egg white, containing whipping improver, 5.8%
Powdered sugar, 20.6%
Salt, 0.3%
Cream of tartar, 0.4%
Water, 45.0%

Formula — second part
Flour, 12.2%
Powdered sugar, 13.5%
Starch, 1.3%
Monocalcium phosphate, 0.5%
Sodium bicarbonate, 0.4%

Procedure

Blend all the dry ingredients in the first part, then add water and whip to a stiff peak. Have the ingredients in the second part mixed separately, sift them, then fold the blend into the whipped egg white, keeping the action to a minimum. Scale the batter into an angel cake pan, and bake at 350°F until the top has browned and the sides begin to separate from the pan. Cool in an inverted position.

Flavors have not been given. Vanilla is always good, though some markets seem to prefer imitation almond extract.

Results

The batter will contain about 6.7% protein, part of it coming from the flour. A two ounce portion of the baked cake will contain about 3.0 to 3.5 grams of good quality protein, about 6% of the daily requirement. In addition, there is no fat (except for about a percent from the flour), no cholesterol, and a low amount of salt. The sugar content is high, of course.

A major source of information on eggs is California Egg Commission, 115 N. Mountain ave. #114, Upland CA 91786. They supply publications relating to eggs, and have developed a computer diskette containing specifications for many egg products. Their phone number is 909-981-4923.

Other animal sources — Protein-containing byproducts of slaughter-house operation are seldom used in foods. Among the byproducts are various forms of blood, bones, trimmings and viscera, hide, hair, feathers, etc. These materials are usually diverted into non-food applications, such as animal feed or fertilizers. In some cases, they may be suitable for food use, if properly processed. An example is gelatin, a widely used ingredient derived from bones or hides. Although such materials might be thought worthy of considation as ingredients for large-scale protein supplementation, their ultimate supply is limited by the number of animals slaughtered for other purposes. And, because they are perishable, they must be either processed quickly or frozen, leading to some economic disadvantages. Changing them into bland protein concentrates for food supplementation may involve techniques not yet invented but probably expensive. Evidently, some blood serum derivatives from cattle are being offered, or have been offered, for food use.

Fish meal, as produced by hot-air drying of mixed varieties of small fish that have not been eviscerated, is used in animal feeds but is not suitable for human consumption. Solvent extraction of fresh whole or cleaned fish can be used to produce a relatively bland high-protein powder that might be accepted for increasing the nutrient quality of bakery foods. The ingredient statement might offer significant problems from a marketing standpoint.

Proteins from Cell Cultures

Much money and labor has been devoted to studying the production of human-utilizable protein by pure cultures of microorganisms. These so-called "single-cell cultures" have included algae, yeast, and molds. Certain of these organisms can turn inorganic sources of nitrogen (such as ammonia or ammonium nitrate) into cell protein when the organisms are grown under appropriate conditions.

The advantage of using single cells is that the organisms can be handled in suspensions, i.e., by liquid bulk-handling techniques, so that growing and processing them is relatively simple and inexpensive. Unfortunately, all (or nearly all) the protein products made in this way have strong flavors. Of all the single-cell preparations that have been suggested, only one has been commercially successful, so far as this author knows. The exception is yeast, some of which is "primary grown" for special supplementation requirements. Inactive dry yeast has an adverse effect on dough properties, probably because of its free glutathione content.

Amino Acid Preparations

The use of protein hydrolysates, consisting mostly of individual amino acids or peptides of short chain length, has been suggested as a means of increasing the PER of bakery products, but their use is attended with some difficulties. For example, Some of the amino acids are inevitably destroyed, regardless of how mild the proteolysis conditions are, so the hydrolysate is never as nutritionally effective as the original protein. And, in the manufacture of these materials, chemical recombinations or rearrangements may occur that could lead to compounds having a deleterious effect on health, even though the original protein was completely wholesome. Flavor and color factors are sometimes undesirable; these may be colors and flavors carried over from the raw material or developed in Maillard reactions between the added amino acids and reducing sugars present in other ingredients.

Certain amino acids are available as almost pure synthetic compounds. Lysine and methionine can be purchased in this form and may be useful for supplementation. They are used extensively in animal feeds, and might be useful in food products the protein content of which could be improved in nutritional quality by remedying a natural deficiency in these amino acids. Improvement of the PER of proteins in foods composed largely of cereals, such as bakery products, is an obvious application of lysine and methionine. Because they tend to be destroyed by heat, it would be advisable to add them to an uncooked adjunct such as an icing or filling if the finished product contains such a component.

These purified amino acids do have some noteworthy drawbacks for supplementation purposes. Among these are high cost and undesirable flavor. The increased cost might be acceptable in certain specialty items, if the consumer can be convinced of the added value, but flavor defects may be more difficult to overcome.

SNACKS CONTAINING MIXTURES OF PROTEIN SOURCES

There have been numerous attempts to develop high protein snacks that avoid the flavor and structural problems of individual supplementing ingredients by combining several sources of protein. The patent literature is replete with inventions of this type. Representative of the attempts is the process described by Ueda and Takaichi (1991) in their patent. One of the examples is described as follows:

Formula (dry ingredients)
Wheat protein powder, 10.6%
Soybean protein powder, 54.0%
Milk protein powder, 10.6%
Skim milk powder, 5.6%
Dry whole egg, 3.5%
Lactic fermentation powder, 3.5%
Water-soluble gelatin, 1.9%
Sweet corn powder, 7.0%
Vitamins and minerals, 2.8%
Vitamin A & D mix, 0.5%

Process
To the pre-blended dry ingredients, 22% of water was added, and the mixture kneaded at 61 rpm in three stages of 15 seconds each. The dough was pressed into sheets using a noodle machine. The dough plates were heated at 120°C in a vacuum of 10 mm Hg for 30 minutes.

Results
The finished product contained 1.6% water, 72.5% protein, 4.5% fat, 13.0% carbohydrates, 0.1% fiber, and 6.6% ash. Due to the vacuum drying/cooking process, the food has a highly porous structure. It has a distinctly short texture, does not seem powdery, is "meltable smoothly in the mouth," and does not adhere to the teeth.

A sweet snack of the cinnamon cookie type was described by Ohren (1972). Formula and processing directions were:

(1) Cream 77.12 parts of shortening with 119.28 parts of granulated sugar.

(2) Add 50.62 parts of liquid whole eggs to the above and beat lightly.

(3) Sift together 100 parts of all-purpose flour, 3.31 parts of cream of tartar, 2.03 parts of sodium bicarbonate, 33.75 parts of cottonseed flour, and 0.81 parts of salt, and add to part (2). Mix well.

(4) Roll the dough into balls of about one-inch diameter. Roll in a mixture of 6 parts sugar and 1 part cinnamon until coated.

(5) Place the balls two inches apart on a greased baking sheet and cook in 400°F oven for about ten minutes.

The cookies contained about 10.6% protein as compared to about 5.6% in a control, made without cottonseed flour. Ohren also described extruded snacks (of the corn curl type) made from a mixture of 80% corn meal and 20% cottonseed meal. The finished corn curl was said to have very good attributes, being yellow, corny, and crisp. It contained about 20% protein vs. 8% in an corn curl made without the cottonseed meal.

A prepared breakfast cereal (similar in many ways to extruded snacks) supplemented with soy isolate was described by Bedenk and Purves (1972). This process is somewhat different from the run-of-the-mill innovations in that it relies on malt to change the other ingredients in such a manner that a superior product is formed, according to the inventors. One of the examples is described as follows. Blend 1,400 parts of corn grits (9% protein), 30 parts salt, and 1,000 parts of water to form a mixture having a moisture content of 41%. Cook this at 250°F under 18 to 20 psig for 60 minutes. Break up lumps by running the cooked mass through a hammer mill. Dry it to obtain a moisture content of 5%. Blend 454 parts of soy isolate (95% protein), 160 parts of sugar, and 57 parts of malt, and 750 parts of water to form a mixture containing 53% moisture; hold this mixture at 120°F for 15 minutes. Mix the corn batch and the soy batch together and extrude the mixture at 550 psig and 200°F. The strands so formed are cut into pellets which are passed through a flaking mill, and then puffed by being contacted with salt held at 320°F. A typical contact time is 10 seconds. The inventors say the flakes are well puffed, tender after mixing with milk, and have a pleasant taste. They have a protein content of 26%. The barley malt spoken of in this patent is an enzymically active ground barley malt, not the syrup.

THE EFFECT OF DIETARY PROTEIN ON BLOOD PRESSURE

A common opinion held for many years among the nutritionist community is that intake of protein above some minimum necessary level tends to increase blood pressure and, in certain cases, impair kidney function. Investigators have found there is a consistent trend to lower blood pressure in vegetarian populations (lower consumption of animal protein, and probably of total protein) than in groups that consume meat on a regular basis. Any conclusions drawn from these correlations are subject to all the many sources of error that shuffling and dealing demographic statistics incur.

A review (Obarzanek et al. 1996) of published studies indicates that this might not be correct; the following discussion is based on their review. The authors mention the "Kempner rice-fruit diet" for patients with severe hypertension, in whom kidney function was frequently impaired, as an example of the application of conventional wisdom. But they point out that studies conducted in Japan suggested that an inverse relationship might exist, i.e., the more protein in the diet, the lower the blood pressure. Data analyses from subsequent observational studies in the US and other countries also provided some evidence of an inverse relationship. For example, a study of over six thousand Japanese-American men in Honolulu reported an inverse relationship between systolic and diastolic blood pressure and dietary protein, as measured in grams per day by a single 24-hour recall. A large study (over ten thousand men and women from 32 countries) showed that protein intake (as measured by analysis of urinary nitrogen and urea nitrogen) was inversely associated with systolic and diastolic blood pressure. Obarzanek et al. report several other studies, all of which tend to support the hypothesis that persons who normally consume more protein tend to have lower blood pressure.

There have not been many controlled studies in which dietary protein was varied over short periods of time and the results of the variations on blood pressure recorded. In publications which were based on such studies, there did not seem to be any direct relationship of dietary protein and blood pressure, and, in many cases, an inverse relationship was observed.

There are not many publications based on animal studies that specifically examined the effects of variations in dietary protein (or amino acids) on blood pressure.

The Committee on Diet and Health, an arm of the National Research Council, stated (Anon. 1989A) that there is no convincing evidence that high protein intake increases the risk of hypertension or stroke and that the role of high dietary protein in renal disease requires more research.

Although these data may seem to be inconclusive, they are indicative, and at least as firmly based in science as the demographic studies upon which most of the dietary recommendations currently in effect have been based. There should be no adverse effects on normal patients of supplementing foods with moderate amounts of good quality protein. Of course, any person who has renal disease or certain other diseases relating to the body's processing of proteins and their end products should follow the directions of their medical adviser, which would usually include instructions not to consume foods that have a high content of nitrogenous substances.

chapter ten

REDUCED SALT AND REDUCED SODIUM PRODUCTS

For years, the public has been bombarded with publicity informing them that salt is bad, that it causes all sorts of physical and mental problems, and that we would all be much better off, spiritually if not physically, if we reduced our consumption of salt, preferably to zero. This is another example of taking a small seed of truth and growing it into a forest of conjecture. Nonetheless, there is a large potential market for reduced salt and no-added-salt products and it is the privilege of the entrepreneur to take advantage of this demand.

INNATE PREFERENCES FOR SALT

The appetite for salt is widespread in the animal kingdom — it is not restricted to humans. This is clearly not a learned response, as some writers would have us believe. Human infants appear to recognize and be attracted by moderately salty flavors in foods. Beauchamp, in an excellent review (1987), detailed the evidence for an innate, genetically determined preference for salt. Much of the following discussion is based on his article.

It appears there are two distinct shifts in the acceptance for salt that occur during infancy. At about four months of age, the child's reaction shifts from indifference to relative preference. Then, at about the third year, there is a change from acceptance to rejection of salt in water while salt is still preferred in other ingestibles. These reactions appear to bear no relationship to the history of exposure of the infant to salt.

The copious anecdotal evidence of animals' preference for salt has been confirmed by many experiments. When a rat experiences a sodium deficiency, it will increase its locomotor activity and become more willing to taste novel substances. When it discovers a source of salt, it will ingest a quantity more than sufficient to replenish its deficiency. This behavior is mediated in large part by the sense of taste, and it is innate — the rat does not need to experience the efficacy of salt in relieving the symptoms of deficiency in order to recognize it as the needed nutrient.

PHYSIOLOGICAL EFFECTS OF SALT AND SODIUM

Sodium is an essential nutrient, although symptoms related to sodium deficiency in the diet are rare because of the widespread presence of sodium in foods. Sodium is required by all animals for a variety of biological functions, including nerve conduction, maintenance of blood pressure at an adequate level, and muscle contraction. Its concentration in the blood is closely regulated; excess sodium is excreted efficiently in the urine, in fact, adult rats fed diets containing 25% salt for three months did not seem to be adversely affected.

Sodium deprivation is quickly followed in many vertebrate species by increases in the activity of the reninangiotensin and in aldosterone production. These effects tend to decrease urinary sodium excretion.

Some people (a minor percentage of the population) are said to have an inherent tendency to high blood pressure that is exacerbated by sodium consumption above certain levels, but there is no evidence that chronic hypertension is initiated by intake of sodium at any level.

The entire rationale behind the idea of improving health and the quality of life and expanding the life span by limiting salt intake is based on interpretations of the results of several demographic studies and numerous experiments with animals and humans that are alleged to show a relationship between hypertension (high blood pressure) and excess salt in the diet. Recently it has been pointed out that the net effect of all this work is inconclusive, so that there is no clear-cut relationship between consumption of salt within normal ranges and hypertension in most persons. There may, however, be a small segment of the population that is genetically programmed to be harmed by intake of sodium at levels normally encountered in American foods, but there are philosophical, ethical, and medical questions as to whether the great bulk of American consumers should be inconvenienced, and have their food choices reduced, and perhaps even have their physiological status unfavorably altered, by reducing the sodium content of all foods available to them in the usual retail outlets.

As an introduction to this controversial subject, several pages will be devoted to an examination of the current state of knowledge of the relationship between sodium and human health.

Causes of Hypertension

The adverse physiological effects of sodium are thought to be expressed primarily in the symptom of high blood pressure. It is very clear that high blood pressure is either the cause of or the result of coronary heart disease and is somehow related to strokes, cardiac failure, and peripheral arterial disease. Hypertension is also the most important risk factor for congestive heart failure, accounting for almost half of the cases.

The principal predictor of cardiovascular problems is the systolic pressure (the highest number in the pair shown on the blood pressure report), the diastolic pressure being less accurate as an indicator of future difficulties.

Because hypertension is a chronic, usually incurable, disorder that causes many serious, disabling symptoms, a great deal of effort has been devoted to finding its causes in order that means for preventing its development can be devised. In spite of many investigations, its etiology is still not entirely clear; there are disputes among investigators as to the relative importance of environmental, genetic, and lifestyle factors.

Some contribution to blood pressure regulation is undoubtedly made by genetic factors; for example, racial differences in susceptibility to hypertension are well known. It is generally accepted, however, that there are at least five causes of blood pressure dysfunction which can be modified: obesity, inadequate physical activity, excess consumption of salt or alcohol, and insufficient potassium intake. There have been suggestions of other factors correlating with hypertension, for example, protein consumption and low-lead.

Although it was originally thought that excess intake of protein led to hypertension (a view going back many decades), a more scientific and dispassionate evaluation of the evidence seems to indicate that there is, in fact, an inverse relationship between dietary protein and blood pressure, i.e., the higher the consumption of protein, the less risk of hypertension developing. Some investigators concluded that vegetable protein was less harmful (if it was harmful) than animal protein, others said no.

Role of the kidneys — Blood pressure and sodium are connected in some manner by events occurring in the kidneys, the latter being a much more complex organ than is generally recognized. The paper of Cowley and Roman (1996) reviews the literature on this subject. A theory attributed to Guyton and associates says that whenever arterial pressure is elevated, activation of pressure natriuresis promotes the excretion of sodium and water until blood volume is reduced enough to return arterial blood pressure to normal levels, i.e., the kidney exerts a negative feedback gain for the long-term regulation of arterial pressure by excreting water (and sodium) thereby adjusting blood volume.

The theory additionally predicts that hypertension can develop only when some agent impairs the ability of the kidney to excrete water (and sodium) and shifts the point of stability toward higher pressures. According to the authors, "This hypothesis in no way presumes that the abnormality leading to the elevation in arterial pressure is intrinsic to the kidney. Indeed, excess sympathetic nerve activity or the release of vasoconstrictor compounds that alter the transmission of pressure to the kidney could readily initiate hypertension." Other possible contributing agents are also discussed in the Cowley and Roman paper. The important point, so far as the present discussion is concerned, is that sodium has not been shown to be either an initial or a precipitating factor in the development of hypertension and its consequent deteriorative effects on the human body.

It is known from experiments with isolated perfused kidney preparations that elevations in the applied pressure increase sodium excretion more or less independently of changes of total renal blood flow or glomerular filtration rate. It seems that pressure natriuresis is mediated by inhibition of tubular resorption in the absence of an apparent intrarenal hemodynamic signal. That is, increases in sodium concentration in the urine result from decreases in the uptake of sodium in the kidney (subsequent to its transfer to the glomerular fluid) rather than increases in the rate of extrusion of sodium ions by the cells.

There is additional evidence that the kidneys are the key to hypertension. When kidneys are transplanted, the recipient takes on the blood pressure characteristics of the donor. Among the examples cited by Cowley and Roman are these: " ... 12 patients who received kidneys from hypertensive donors had higher blood pressures and required more aggressive hypertensive therapy than 11 matched patients who received kidneys from normotensive donors," and " ... transplantation of a kidney from a normotensive donor produced a permanent normalization of blood pressure in six hypertensive patients with long-standing essential hypertension that eventually had led to end-stage renal disease."

The picture drawn by Cowley and Roman becomes extremely complicated, too much so to justify inclusion in the present discussion, but the net result, in your author's opinion, is that sodium levels in the blood can be adjusted over a large range, and are so adjusted, by events occurring in the kidney and elsewhere, and in persons with efficiently functioning kidneys, a considerable excess in sodium intake can be accommodated and "corrected" by actions occurring in the kidneys and perhaps elsewhere. with no more than a transient effect on blood pressure, if that.

Even if it could be clearly shown that there is a malfunction, or an inadequate degree of functioning, of the renal-related excretion and uptake of sodium (and possibly of other substances) in persons who are susceptible to the development of dangerous increases in blood pressure, it may be that this malfunctioning is not restricted to the control of sodium concentration in the blood and that there may be other effects much more directly related to blood pressure control.

The role of sodium – A direct connection, a causal relationship, between sodium concentration and coronary heart disease has not been clearly established, leaving us with the thought that the problem may be much more complex, and much less of a simple cause and effect relationship, than has been recognized heretofore. In other words, defective handling of sodium transfers across the membranes in the kidney may be accompanied by hypertension, but there is no clear showing, at this time, that one causes the other – they may both be caused by another factor that is receiving inadequate attention.

In a recent meta-analysis of randomized controlled trials (Midgley et al. 1996), covering 56 published studies in which the effect of reduced dietary sodium on blood pressure, was examined, it was found that there was "significant heterogeneity" in the results. Only trials that included randomized allocation to control and dietary intervention groups, monitored by timed sodium excretion, and with outcome measures of both systolic and diastolic blood pressure were selected by blinded review (i.e., the reviewer was not informed of the identity of researchers or their employer, etc.) of the methods section of the basic publications. The conclusion of the authors was "Dietary sodium restriction for older hypertensive individuals might be considered, but the evidence in the normotensive population does not support current recommendations for universal dietary sodium restriction." This conclusion was based on decreases of about 3.7 mm Hg for systolic pressure in hypertensives, and about 1.0 mm Hg in normotensive individuals, adjusted to reflect a reduction in daily excretion of sodium (an indication of sodium intake) of 100 mmol (equivalent to somewhat less than 6 grams of salt). The decreases in blood pressure were larger in trials of older hypertensive individuals and small and insignificant in trials of normotensive individuals whose meals were prepared and who lived outside an institutional setting.

An interesting possibility suggested by the statistical analyses of Midgeley et al. is that small trials showing *no effect* of dietary sodium restriction on blood pressure were underreported, although they say, "It is also possible that the exaggerated treatment effect [i.e., results of sodium restriction on blood pressure were overemphasized by the researchers] was related to poorer methodological quality, which appeared to be more common in smaller trials."

Many nutritionists do not want to believe that there is little or no health-improving effect of dietary salt restriction. Lenfant (1996), who is on the editorial staff of the Journal of the American Medical Association, states, partly in response to the Midgely et al. article discussed above, "Regardless, the preponderance of evidence continues to indicate that modest reduction of sodium, as recommended in the 1995 US Dietary Guidelines for Americans, would improve public health." It would be interesting to know where the "preponderance" can be found; possibly in the minds of the people who conspired to establish the Dietary Guidelines.

Other known factors in hypertension — Factors other than salt intake are known to be associated with hypertension and the numerous illnesses that seem to radiate from that point d'appui. Obesity is one of these, of course, but also implicated are lipid abnormalities and glucose intolerance, though they are not as clearly connected to the onset of hypertension as is grossly excessive overweight.

A final solution — In spite of the uncertainty in demographic and experimental data, the nutritionists at the FDA, in their infinite wisdom, have been able to determine that the daily reference value for sodium will be 2,400 mg, equal to the amount of this element in about six grams of salt. The average daily sodium intake of American adults has been calculated at about 3,900 mg, equivalent to about 10 grams of salt.

Is It Salt or Is It Sodium?

When we discuss dietary implications of salt, we are really thinking about sodium ion. The other element in common table salt — chlorine — is seldom considered, although it, too, has interesting effects on human physiology, being in fact an element essential to human survival.

As stated above, many researchers have concluded, rightly or wrongly, that ingesting sodium chloride (above a certain low base level) will cause elevation of blood pressure in some individuals with hypertension; it follows that reducing the amount of salt consumed by these persons will reduce their blood pressure. This is a current tenet in regulatory circles.

It has been more or less taken for granted that the sodium ion is the culprit, but a study published in 1984 reported that salt-sensitive rats responded to sodium chloride intake with a rise in blood prssure, while an equimolar intake of sodium with anions other than chloride did not cause hypertension. Adequate chloride was provided for growth of these animals, and the absence of hypertension despite high levels of sodium suggested that a combination of sodium and chloride is necessary for the hypertensive effect.

In a study involving hypertensive human subjects whose blood pressure returned to normal when fed a low sodium diet (i.e., they were the salt-responsive type of hypertensives), blood pressure rose when the diet was supplemented with sodium chloride but not when it was supplemented with sodium citrate (these studies reviewed in Anon. 1988C). A series of studies attempting to correlate calcium intake (and sodium and calcium combinations) reviewed by Weaver (1988), led to conflicting conclusions, but the preponderance of data seems to favor a role (perhaps a minor or indirect one) cf calcium in blood pressure control.

A Minority View

These results may lead you to think that the relationship of sodium and/or salt to hypertension is more complicated than your doctor tells you. It is interesting to speculate on the physiological properties of the halides: chlorides, bromides, iodides, and fluorides. For more than a hundred years, a mixture of bromides (often ammonium, potassium, and sodium bromides) has been used as a sedative. Is it possible that their calming effect, which might be considered the reverse of the hypertension symptom, is due to a displacement of chlorine by bromine at a bioreceptor. Iodine is known to be an essential participant in functions of the thyroid gland, which has pronounced effects on mental as well as physical well-being. Fluoride is a virulent cell poison.

Of course, some of the alkali metal (lithium, sodium, potassium, rubidium, and cesium) cations are also known to have important effects on systemic tension; the beneficial effect of lithium on the mood swings of manic-depressives is well known. There may well be a complex interplay of these cations and anions in controlling smooth muscle tension that extends far beyond blood pressure symptoms and which has yet to be elucidated.

REALITIES OF THE MARKETPLACE

Regardless of the basic factors affecting hypertension, many people select at least some of their foods on the basis of the product's salt and/or sodium content, giving preference to products that claim "No salt added," "Reduced sodium," or the like, provided other factors (price, taste, etc.) are the same. Suspicion of sodium has become deeply embedded in the minds of many consumers. We have read letters-to-the-editor from persons that hysterically upbraid restaurant managers (in general) for making available to their patrons that poisonous white powder (salt). However illogical the aversion is, a demand does exist that can be satisfied by offering low-sodium, no-sodium, and no-added-salt foods, and many food manufacturing companies are attempting to satisfy that demand.

PROBLEMS AND POSSIBILITIES

Imitating the desirable taste resulting from the use of salt as an ingredient is, of course, the predominant goal of any project having as its purpose the development of products having reduced or zero content of sodium chloride. There are other functions of salt, however, that must be considered when designing these products. Salt — or rather its ions sodium and chloride — have physical and chemical effects on other ingredients that cannot be ignored if intermediates having processing characteristics adaptable to available equipment and finished products having the texture and appearance desired by consumers are to be achieved.

A matter of good taste

In examining the sodium content of commercial loaf bread, DuBois et al. (1984) found from 266 to 360 mg per two ounce serving, in a series including white pan bread, wheat bread, whole wheat bread, French/Vienna bread, and cracked wheat bread. It is reasonable to suppose that these levels represent long experience with the demands of the consumers being served by the bakeries. Arbitrary reductions in salt content would entail a risk of losing market share.

If taste considerations can be ignored, it is fairly easy to formulate or re-formulate most bakery products in a manner that will justify low sodium claims (or, a claim of "no added salt"), but some differences in fermentation rate and dough properties will result. These problems can be offset, usually, by modifications in procedures. But, the low-salt products will almost certainly have lower acceptability ratings than their conventionally formulated counterparts. Simply stated, they do not taste as good.

The less attractive flavor may be accepted as unavoidable by the sodium-sensitive customer, but the remainder of the consumer population will probably seek out an alternative product. If the consumer who is not concerned about salt is turned off by a dietetic product from manufacturer X, he may avoid X's entire product line in the future.

Although it is not essential to make a label claim of reduced sodium or reduced salt, even for a standardized product, if the salt content is reduced to a very low level, it would seem to be important for the manufacturer to clearly indicate the special dietetic nature of salt-free products so that they are not inadvertently chosen by a consumer who is under the impression they are standard (high-salt) products. There is also the possibility that consumers may not read the label, other than the brand name, and not understand they have bought a low-salt product. This may suggest the need for selecting a totally different brand name and label design for the dietetic products.

Salt Substitutes

Replacements for salt have been sought as a basis for formulating low-salt or no-salt products that might appeal to persons with health problems (particularly those afflicted with high blood pressure), but none of them completely reproduces the flavor effects of salt. Among the additives that have been offered as salt replacements are spices, vinegar and other acidic ingredients, and potassium chloride. The latter substance has some of the salt flavor notes, but it may not have the flavor intensifying effect of salt, and it seems to have some flavor and mouthfeel characteristics peculiar to itself.

Research work is being done on putting to work the salty taste of certain peptides. Japanese scientists have pointed out that ornithyltaurine monochloride and ornithyl-β-alanine monochloride taste salty. If the HCl is not present, the salty taste is not observed, and in general lowering the pH of the sample increases the saltiness (up to a point). Evidently, these compounds taste sour, but adding some sodium chloride can mask this acidic note. The scientists say that sodium ion consumption can be cut 75% while keeping a salty taste (Worthy 1990). So far as can be determined, these substances do not have FDA approval for use as food ingredients. Even if permitted, it is clear the potential cost of these ingredients would deter all but the most sanguine merchandisers.

The importance of salty flavor varies with consumers' expectations, and their expectations are generally based on experience. For example, salt is the most important part of the flavor complex of nearly all snacks. Even if all other characteristics of corn curls or potato chips could be kept constant, the absence of salt would be noticed by the consumer and, in nearly all cases, his evaluation of the desirability of the product would be unfavorable. A slight increase in sweetness of the product may partially mask the absence of sodium but, of course, sweetness is not a freely manipulatable dimension in many products, Spices have also been used to compensate for the missing ingredient but they always add a characteristic flavor of their own that is rarely expected and seldom appreciated by the average consumer.

Other ingredients which have been proposed as a means of diverting the consumer's attention from the lack of salt flavor include MSG, meat extracts, protein hydrolysates, autolyzed yeast, soy sauces, and nucleic acid derivatives.

Flavor Problems with Salt Replacers

A possible complication in any program to develop foods having reduced salt content is that the improving effect of salt on flavor is not due solely to the taste experienced when, for example, salt is mixed with water and then sipped. In compounded foods, salt seems to modify the tastes of other components, enhancing the perception of some flavors and decreasing it in others. Using a substitute for salt that does not reproduce all of the effects of the ingredient on the flavor profile of the finished product risks endangering the consumer acceptance of the food.

Potassium chloride — The most widely promoted, and probably the most widely used salt substitute is potassium chloride. This substance does have a somewhat salt-like flavor, but it also contributes off-flavors, variously described as metallic, bitter, chemical, etc. Some authorities say the bitterness of potassium chloride becomes a consumer negative only when it is used in certain products, such as soups and yoghurts. Furthermore, it is claimed that some spices, particularly Italian-type spice mixtures, will mask the off-notes of potassium chloride.

Employing mixtures of salt and potassium chloride is another approach to ameliorating the taste problems, and a 50:50 mixture of these two ingredients is commonly used with reportedly good results in reduced-sodium foods. A mixture suggested by Adams (1983) contained approximately equal portions of sodium chloride and potassium chloride, with traces of fumaric acid, potassium glutamate, and monocalcium phosphate. When used at about the 2% level in bread doughs, this blend gave a well-flavored but rather bland bread containing only 100 mg of sodium per two-ounce serving. Some processing responses of the dough were affected. Additional yeast and oxidants were required when using the preferment process and about 30 minutes additional fermentation time was required when using sponge-dough processes. The bitter note of potassium chloride is definitely detectible in bread containing a significant amount of this seasoner.

If a chemically leavened product is being made, substantial amounts of sodium will be brought into the product by the sodium bicarbonate and probably by the leavening acids, and these sources must be taken into account when developing labeling and advertising claims.

Gillette (1985) correctly pointed out that flavor studies conducted with aqueous solutions of sodium chloride or its substitutes give results that are of limited use to food technologists. When her panels tested soups, rice, potato chips, and scrambled eggs made with various levels of salt and salt substitutes, they reported that the addition of sodium chloride did more than just add a "salt" flavor. Salt increased the perception of "fullness" and "thickness," giving the impression of a less watery product. It enhanced the perception of sweetness. In some cases, metallic or chemical off-flavor was decreased or masked by salt. Overall, the most significant effect of sodium chloride was its improvement of the flavor balance. These effects were often obtained even when "saltiness" per se, was not noticeable.

In the Gillette studies, potassium chloride enhanced sweetness but did not improve mouthfeel or flavor balance significantly. It did, however, impart bitter, metallic, and chemical notes to products. Glutamate also lacked the overall improving effect of sodium chloride. The net effect of all this experimentation is to confirm something that most of us already knew, that is, many food products taste better if they contain some salt. Many persons who believe they must avoid sodium can learn to accept salt-free or reduced salt products, but like most other educational programs, the learning is accompanied by considerable discomfort and the end result may be worthless.

The perceived saltiness of sodium salts varies with the anion present. With anions other than chloride, the intensity of flavor changes, and tastes reported as "off-flavors" often occur. Ye et al. (1991) investigated the mechanism of this phenomenon and concluded that field potentials across the lingual epithelium modulate taste reception, so that the functional unit of taste reception includes the taste cell and its paracellular environment.

Table 10.1
TYPICAL PROPERTIES OF COMMERCIAL POTASSIUM CHLORIDE[1]

Characteristic	USP limits	Code 6838 granular	Code 6842 powder
Assay, % KCl	99.0-100.5	99.6	99.6
Sodium, % max.	to pass	0.004	0.006
Heavy metals, %Pb max.	0.001	<0.001	<0.001
Arsenic, % max.	0.0003	<0.0003	<0.0003
Moisture, % max.	1.0	0.04	0.04
Particle size, % on sieve:			
No. 20	--	0	0
No. 30	--	0.5	--
No. 40	--	17	trace
No. 60	--	55	18
No. 70	--	7.5	16
No. 100	--	14	33
thru No. 100	--	6	33
Loose density, lb/cu.ft.	--	67	65
Loose density, fl.oz/lb	--	14	15
Tapped density, lb/cu.ft.	--	70	70
Tapped density, fl.oz/lb	--	14	14
Solubility rate[2]	--	54	28

[1] The "Code" numbers identify brands of Mallinckrodt, Inc.
[2] Solubility rate is expressed in seconds for 30 grams KCl in 100 grams agitated water at 45°C.
NOTES: Also, rated as "Passes" for iodide or bromide, acidity or alkalinity, and Ca and Mg.

Effects of Salt Reduction on Bakery Processes

The concentration of salt in doughs, batters, and snack mixes affects a number of processing responses — some favorably, some unfavorably. In the production of yeast-leavened bread and rolls, salt acts as a regulator of fermentation and affects gluten strength. These functions make make it more difficult to process a salt-free bread — it is not a matter of simply deleting this ingredient from the formula while keeping all other factors constant.

Salt is commonly added to bakery formulas at levels of from 1% to 2.5% of the flour weight, with more of the formulas probably specifying nearer the lower than the upper part of the range. In addition to having salt in the dough, some specialty products such as saltine crackers, salt sticks, and pretzels carry an even larger amount of salt as sprinklings of fairly large particles on their surface. Placing the granules on the surface has two beneficial effects: an immediate flavor impact is obtained when a piece of the product is placed in the mouth and the salt does not affect dough properties.

Since salt must first dissolve in order to be detected by the taste buds, the particle size of topping salt and its dissolving rate is a direct function of its exposed surface area, which is related, in a general but inexact way, to the particle size. Given the same crystal form, surface area (per weight) increases geometrically with decreasing particle size, so that salt sample X with a particle diameter one-half that of sample Y would, initially, dissolve about twice as fast as the coarser grade, and would give twice as much flavor impact. In practice, this rule seems to apply fairly well to different sizes of the coarse topping salt used with crackers and pretzels, but not as well to finer toppings employed on chips, nuts, and popcorn (Strietelmeier 1974).

When salt is replaced with potassium chloride to make a reduced sodium bread, problems in addition to flavor changes will be encountered. Sodium chloride affects both the fermentation rate of yeast and the rheology of the dough. Salt "controls" fermentation through its inhibitory effect on fermentation. This effect is not due solely to the increased osmotic pressure, but is partly a specific manifestation of the action of sodium and chloride ions on the semipermeable membranes of yeast cells. Yeast does not respond to potassium chloride in the same way it responds to sodium chloride, so the baker will have to establish new times for fermentation and proof periods if salt is to be taken out of the formula and potassium chloride added.

Although salt affects many of the metabolic activities of yeast, the effect on fermentation that is of the most importance to the baker is the reduced rate of gas production and the consequent lengthening of proof time. Data in the literature indicate that, at least for some doughs, the reduction in gas production rate is approximately linearly related to salt concentration. Therefore, in reduced-sodium doughs, the dough pieces will expand more rapidly, provided temperature is kept the same. Although this might be regarded as a favorable development, leading to increased throughput, the possibility of wild fermentations that cause the dough to overflow the pans and cause other problems must be considered. Additional problems are the possible exhaustion of fermentation substrates in the intermediate fermentation step so that adequate expansion cannot be obtained during pan-proofing. Also, excessive fermentative activity due to inadequate salt in the dough may leading to sour, off-flavored doughs that are difficult to process, and which usually result in baked loaves that have open grain, poor texture, and other defects.

Salt has a modifying effect on the physical properties of gluten. This is usually described by bakers as a binding or tightening effect. Increases in salt lengthen the mixing time — the more salt, the longer the dough must be mixed in order to achieve proper development. Furthermore, the energy required for adequate mixing may be greater. Less salt leads to softer, stickier doughs that tend to hang up in the machinery.

Some experiments conducted several decades ago led the researchers to attribute the effect of salt on gluten to an inhibition of proteolytic activity. If this does occur, it must be a relatively minor part of the effect, however. More likely, the sodium ions compete with hydrogen ions for loci on the gluten molecule, reducing the hydrogen bonding between the molecules. Fortmann (1967) sees the phenomenon as a coagulation or hardening of the gluten with a release of water; the gluten molecules become bent and inter-locked so that work in the form of additional mixing is required to unwind and unravel the gluten strands into more nearly linear alignments.

As a result of its effect on dough, salt generally improves the crumb color and grain in bread. That is not to say that conditions cannot be adjusted so as to get a loaf of nearly equivalent quality in salt-free bread if the conditions and other ingredients are adjusted correctly — in fact there is inadequate information on which to base a statement of this sort. Also, data on the effect of potassium ion on dough and bread quality appears to be lacking in the published literature.

A formula and process for white pan bread made by the short time process and having the salt content reduced to 1% (flour weight basis, 0.56% dough weight basis, probably slightly less than 0.7% finished product basis) was published by DuBois et al (1984)

Formula

100.0 parts bread flour
  6.0 parts granulated sugar
  1.0 parts salt
  3.0 parts shortening
  3.5 parts compressed yeast
  0.5 parts yeast food, acid type
 67,0 parts water
    10 ppm potassium bromate

Procedure

1. Mix dough on low speed 1 minute, and to development on second speed. Dough to come out at 83°F.
2. Ferment 30 minutes at 84°F.
3. Divide, scaling 17.5 ounces each.
4. Give an intermediate proof of 12 minutes at 77°F.
5. Sheet, mold, place in pan.
6. Proof to 0.625 inches above the top of the pan at 105°F.
7. Bake 10 minutes at 420°F.

Results

Rate of gas development increased over a control loaf, necessitating a decrease in proofing time. When proofed to a standard height, loaf volume was affected only slightly. Bread quality improved by addition of salt. In a series of salt variations, an untrained panel could not consistently detect a flavor change between 1.7% and 2.1% salt (FWB); below this level, the number of panelists correctly identifying the flavor change increased, and the number preferring the higher level of salt also increased.

Sodium is present in many food ingredients besides common salt. One of these ingredients is sodium bicarbonate (baking soda), the essential leavening component (gas source) in thousands of formulas for chemically leavened baked goods. The formulator who desires to develop a replacement for a traditional high-sodium bakery food will encounter may problems, flavor being but one of them. However, removal of sodium ion from cake and muffin batters, cookie doughs, pie crusts, and the like can be expected to have less of an impact on the physical properties of the intermediates and on the texture and appearance of the finished products than in yeast-leavened doughs because gluten development, an essential part of the preparation of breads and rolls and is not an important feature of the processes of making batters.

Other approaches — We would have to place the invention of Lee and Tandy (1996) in the miscellaneous category. Their patent covers the process of encapsulating "ammonium salt" so that it can be added to a food or beverage without causing some of the adverse effects normally expected of such compounds, yet would enhance or potentiate the salty taste of the product. The salt taste enhancers covered by the patent are not claimed as salt substitutes and cannot completely replace salt, but are intended to allow foods and beverages to be formulated with lower amounts of salt.

Replacing Sodium Bicarbonate

As mentioned previously, the great majority of chemically leavened baked goods rely on sodium bicarbonate as the source of gas. The acid-reacting component of the leavening system is often, but not always, another sodium salt. To reduce the sodium content of the final product to a minimum, it is necessary to use sodium-free chemicals for both the compounds in the system. Potassium carbonate, potassium bicarbonate, and calcium carbonates are obvious candidates for replacing baking soda, and numerous publications have suggested methods by which these compounds could be employed to make low-sodium cakes, cookies, pancakes, biscuits, and the like.

Many old and expired patents dealt with the use of ammonium carbonate, ammonium bicarbonate, calcium carbonate, and other carbonate or bicarbonate salts as the source of carbon dioxide in leavened doughs and batters. Today, ammonium bicarbonate is occasionally used as part of the leavening system in certain cookies and crackers, where its ability to cause the cookie doughs to spread out and the surfaces to smooth is regarded favorably.

Verduin (1991) patented a process for making low sodium sponge goods, which, though yeast-fermented, normally require the addition of sodium bicarbonate in a late stage in dough processing; this class includes such familiar items as soda crackers, oyster crackers, pretzels, and flavored snack crackers. Her process uses potassium carbonate and potassium bicarbonate as part of the leavening system. Since these materials replace sodium bicarbonate that would otherwise be used as a source of carbon dioxide, it is apparent that foods prepared by the patented method would be lower in sodium content than crackers, etc., prepared by conventional formulas and methods. Advantages of the disclosed invention, as compared to prior art, are said to include a substantially uniform texture and pH throughout, a substantially uniform surface color, and pleasant mouthfeel and taste.

An example provided in the patent of Verduin includes the following formula and process.

Sponge formulas

64.00 parts flour
0.20 parts yeast
0.46 parts malt flour
1.5 parts sponge meal
25.0 parts water

Sponge process
The flour was sifted, then mixed with the malt flour and sponge meal. The yeast was dissolved in the water, allowed to stand for at least 10 minutes at room temperature, then mixed with with remainder of the ingredients. The dough container was covered and allowed to ferment for 19 hours.

Dough formula
To the sponge, made as previously described, add:
36.0 parts flour
8.0 parts shortening
2.42 minor ingredients (salt, malt extract, inactive yeast, calcium carbonate)
0.25 parts potassium bicarbonate
0.75 parts anhydrous potassium carbonate

Processing the dough
The potassium carbonate and the potassium bicarbonate were mixed and dissolved in a portion of the water, which was then added to the sponge. The dough flour, then the salt, calcium carbonate, yeast, and malt extract were added to the trough, which was then placed in the spindle mixer. The mixer blades were started, and the shortening added while mixing. After mixing was completed, the dough was proofed for four hours, then it was laminated and cut. Topping salt was applied by a salt spreader at the rate of 1.38 parts by weight. After baking, spray oil can be applied.

Results
The crackers resulting from the patented process, in which one-third of the salt was present as dough salt and two-thirds as topping salt, had a taste substantially the same as the taste of control crackers made in the same way but with sodium bicarbonate. However, the experimental crackers has only 115 mg of sodium per serving, compared to 190 mg per serving for the control crackers. The experimental formula crackers had a substantially uniform pleasant light tan color. Dark streaks, which are indications of potassium hot spots, were not present, and the crackers had a substantially uniform laminar cell structure throughout and a pleasant mouth feel, pleasant crispness, and no bitter off-flavor.

Although some of the acid-reacting components of chemical leavening systems do not include significant amounts of sodium ions, others, such as sodium acid pyrophosphate do contain several percent of sodium. Replacement of these leavening "acids" can pose reformulation problems that may be difficult to solve, since the sodium salt often acts as a buffer and has been designed to facilitate release of carbon dioxide in a timed pattern. Various attempts have been made to develop improved non-sodium leavening acids. The patent of Heidolph and Gard (1995) describes the use of calcium acid pyrophosphate, which causes the delayed gas release in bicarbonate-containing doughs that is characteristic of sodium acid pyrophosphate. In a dough-rate-of-reaction-test, it exhibits a maximum rate of release of carbon dioxide between 17 and 40 minutes after initial mixing.

Making a Good First Impression

When a food is taken into the mouth as a piece, rather than as a fluid, the full effect of the salt distributed throughout the piece is usually not sensed. Although the pieces may be broken down considerably by chewing so that the particles become quite small, it is certain that some at least of the ingredient salt in the interior of the particles will never reach the sodium receptors in the mouth. It seems reasonable that the flavor impact of a given amount of salt could be magnified by concentrating the added salt on the surface of the food pieces. In this way, a smaller amount of salt can be made to be perceived as a higher concentration by placing it on the surfaces immediately contacted by the saliva. To be effective, it is obvious that the particles cannot have an continuous aqueous phase throughout, for then the salt, wherever it is placed, would eventually diffuse throughout the fluid system in the days or weeks before the product reaches the consumers.

Particularly suitable for this kind of "salt-maximizing" strategy are savory snacks. Salt placed on the surface of popcorn, potato chips, and crackers is likely to be as effective as several times its weight of salt distributed uniformly throughout the pieces. Breakfast cereals may or may not be modified in this way; if the customary method of use is to place the dry pieces, such as corn flakes, in a bowl of milk, then the salt on the surface will first distribute itself throughout the external fluid phase and eventually throughout the interstices of the particles as the fluid penetrates into the particles' interiors.

The patent of Fan (1991) disclosed a method for making a ready-to-eat cereal having good initial taste in spite of reduced sodium content. In essence, this patent covers a better way to apply salt to the surface of breakfast cereal pieces by spraying on them a concentrated salt solution. The ratio of salt applied by spray and the salt in the product formula is the critical point. After the spray has been applied, the cereal is dried to about 5% moisture content. Tests by sensory panels showed that cereal prepared by this method received high scores for taste dimensions, including saltiness.

chapter eleven

ALLERGIES AND OTHER FOOD SENSITIVITIES

Nutrients — that is, the supply of energy and the chemical building blocks that are used by the body to make tissues and support metabolism — are not the only substances ingested in the form of food. Accompanying the useful compounds are other substances that are either innocuous (like water) or harmful. A frequent observed characteristic of the latter is the wide variation in susceptibility and extent of damage suffered by different people. The reason for these differences is not always known; sometimes it appears to have a genetic basis, at other times it seems that a chance prior encounter with activating (or deactivating) substances is the cause.

Types of Adverse Reactions

One system of classification of the causes of harmful reactions to food components includes the following categories:
Primary food sensitivity:
    Non-immunological:
        Allergy-like intoxications
        Anaphylactoid reactions
        Metabolic reactions
        Idiosyncrasies
    Immunological/Food allergies:
        Non-IgE-mediated
        IgE-mediated/Food anaphylaxis
            Exercise-induced
Secondary food sensitivity:
    Secondary to GI disorders
    Secondary to drug treatment
Inability to normally process certain foods:
    Lactase deficiency
    Chinese restaurant syndrome
    Migraine triggered by ingestables such as red wine and aged cheese.

The formulator of foods for the general market can hardly be expected to take into consideration all of these idiosyncrasies, some of which affect only a few individuals and others that cause only minor and transient symptoms, but there are opportunities to develop and supply dietetic versions of normal products that fit the needs of enough sufferers to constitute a viable market segment. The prevalence of food allergies in the overall population is said to be less than 2%, the frequency being higher in infancy and diminishing with increasing age.

Terms Used to Describe Adverse Reactions

As in most scientific disciplines, allergy science has acquired a specialized vocabulary which has to be understood before much sense can be made of published accounts on the subject. Following are a few of the terms that will be encountered in the present discussion.

adverse reaction – a clinically abnormal response attributed to exposure to food.

anaphylactic reaction — a reaction that mimics hypersensitivity reactions; there is no immune response.

food allergy (hypersensitivity) — immunologic reaction resulting from exposure to some kind of food; reactions occur only in some subjects; reaction is apparently unrelated to physiological effects of the food.

food idiosyncrasy — quantitatively abnormal response to food; reaction differs from its pharmacological and physiological effects and resembles hypersensitivity reactions; there is no immune reactivity.

food intolerance — abnormal physiologic response to food; the reaction is not immunologic.

food toxicity (poisoning) — adverse effect caused by direct action of food; mechanism is not immunologic; toxins may be contained within the food or released by microorganisms contaminating the food.

metabolic reaction — a reaction that results from the effect of the food on the metabolism of the host.

pharmacologic reaction — an adverse reaction to food as a result of a naturally derived or added chemical that produces a drug-like effect in the host.

ADVERSE REACTIONS THAT ARE NOT ALLERGIC

A indicated in the review of Hefle (1996), food sensitivities can be defined as reproducible, unpleasant reactions to a specific food or food ingredient; in extreme cases, the reaction can be life-threatening. The mechanisms for some of these syndromes, such as the Chinese restaurant syndrome and food (and beverage) associated migraine, are unknown. Although the public tends to call all of these unpleasant response to foods "allergies," many of them do not have the essential characteristics of true allergies.

Intolerances to food that do not involve the immune system, and thus are not truly allergies, are often difficult to diagnose and correct. Indeed, many of them seem to have a strong psychological component and cannot be remedied by treatments that neglect this consideration. Metcalfe (1992) the following possibilities (other than allergies) must be considered when making a differential diagnosis of an adverse reaction to a food:

1. Gastrointestinal disease [e.g., inflammatory bowel disease]
2. Enzyme deficiencies [e.g., sucrase deficiency]
3. Additives and contaminants [e.g., dyes and exogenous chemicals]
4. Toxins [e.g., bacterial and fungal toxins]
5. Endogenous chemicals [e.g., histamine, alcohol]
6. Psychological reactions
7. Systemic disorders [e.g., endocrine disorders]

Presumably, any of these disorders could be mistakenly regarded by the patient (at least for a time) as an allergic reaction.

Lactose Intolerance

According to one report, between 30 and 50 million of Americans may suffer from lactose intolerance (clinically known as lactase non-persistence). Certain ethnic groups tend to be affected more than others: blacks and some Asians are particularly likely to have some degree of lactase insufficiency. Estimates of the prevalence of lactose intolerance in various ethnic groups have wide ranges, see the table below (Best 1993):

Caucasian, 6 to 25%
African American, 45 to 81%
Mexican American, 47 to 74%
Asian American, 65 to 100%
Native American, 50 to 75%

Lactose is present in the milk of all mammals but rarely, if ever, is synthesized by other organisms. Therefore, the symptoms of lactose intolerance can be elicited only by some form of milk and, apparently, by milk from any source — human, cow, goat, etc.

Lactose intolerance is manifested by intestinal discomfort and allied symptoms after more than a minor amount of milk has been ingested. It is said to be related to inadequate production of lactase, the enzyme that breaks down milk sugar (lactose) into glucose and galactose. These two monosaccharides are readily absorbed from the digestive tract and cause no metabolic difficulties, but non-hydrolyzed lactose passes into the lower intestine where it is used by the microorganisms normally present there. Gas and acid are produced as byproducts of the bacteria's metabolism, and these materials are largely responsible for the unpleasant symptoms.

Lactose intolerance has been diagnosed by dosing the suspect with as much as 50 grams of lactose — about the amount found in a quart of milk. After a time, the breath of the patient is tested for hydrogen gas. A high level of hydrogen indicates inability to digest lactose. The hydrogen is produced in the lower intestinal tract by microorganisms that feed on the undigested lactose that reaches them; it enters the blood stream, is carried to the lungs, and then transferred to the exhaled gases.

Lactase insufficiency is seldom found in infants, they are genetically programmed to fully utilize milk since that is the food on which they survive for the first part of their life.

Lactose intolerance must be distinguished from allergy to milk proteins, which has an entirely different etiology and which may be restricted to certain kinds of milk, i.e., it may appear as a symptom in infants ingesting cow's milk who can safely consume human milk. Furthermore, very small amounts of milk can trigger an allergic attack, while the symptoms of lactose intolerance are strongly dependent on the quantity of milk consumed, many so-called lactose intolerant subjects being able to drink a cup or so of milk with little or no discomfort.

Preventing the symptoms — A certain proportion of the adult population is unconcerned about lactose intolerance. They don't drink milk and don't intend to drink milk. The small amount of lactose consumed as part of cheese, ice cream, whipped creams, coffee cream, etc., is almost always insufficient to trigger the reaction, so they are never made aware of their inability to efficiently utilize the sugar. If discomfort is occasionally felt, they attribute it to some other factor or ignore it. There is however, a fairly large segment of the population that feels the need for milk as a beverage, or who are so sensitive to lactose that even small amounts of the sugar can cause adverse reactions, and for these a remedy would be welcome.

Dairy product manufacturers have, of course, been anxious to develop modified milks that can be consumed with comfort by lactose intolerant persons. A straightforward way of doing this is to remove the lactose from regular milk, or at least to reduce the lactose content to a very low level. This can be done by various separation techniques, such as membrane dialysis. Another approach is to add enzymes to break down the lactose into its constituent monosaccharides. McNeil Consumer Products division of Johnson and Johnson began nationwide marketing of an UHT-processed, lactose-reduced whole milk in 1980, and has since introduced Lactaid-100, a 100%-lactose free nonfat milk. Other companies have since entered the market with similar products. All of these sell at a substantial premium to regular milk. The government requires a 70% reduction in lactose content (compared to regular milk) before a lactose-reduced label claim can be made. Hydrolysis greater than this can cause noticeable, and often objectionable sweetness — the glucose and galactose formed by splitting lactose contribute substantially greater sweetness than the disaccharide from which they were formed.

Tablets and liquids containing a form of lactase are being marketed so the consumer can add them to milk, or take them with milk, in order to avoid the unpleasant symptoms he attributes to lactose intolerance.

Formulated milk products, in which are combined milk proteins or whey proteins, a non-lactose carbohydrate (usually sugar or corn syrup solids), flavors, and often some kind of fat, have appeared on the market with labels indicating their suitability for consumption by lactose intolerants. Labeling is a problem, since most if not all of these products cannot be called any kind of milk except imitation milk. Milk replacements that do not contain any milk-derived ingredient not only meet the needs of lactose-intolerant persons but might also find a market among kosher consumers who avoid mixing dairy and non-dairy foods.

Another approach, possibly not yet commercialized, is to add live yoghurt cultures to milk. The theory is that the bacteria stay dormant at refrigerator temperatures, but become active when they reach the lower intestine and feed on the lactose. It is not clear why this would prevent normal bowel flora from also attacking the lactose in competition with the implanted bacteria, thus causing the usual problem, though perhaps in a less severe form.

At least one manufacturer (Diehl Specialties International) produces liquid and powder forms of a lactose-free milk substitute concentrate that is distributed to dairies where it is reconstituted, packaged, and sold as a dairy drink. The product, tradenamed "Vitamite," contains canola oil, and some soy protein, and is fortified with calcium, vitamins B-12, A, D, and B-6. and vitamins

In 1996, the FDA affirmed that lactase enzyme preparation from *Candida pseudotropicalis* is GRAS for use in milk-derived products to hydrolyze lactose (61 Federal Register 7702-7704).

Formulated products without milk — The development of bakery products suitable for lactose intolerant consumers should be a relatively simple task, at least for most varieties of these foods. Satisfctory breads and bread rolls can be made without any dairy ingredient — it is being done every day. This also applies to most kinds of doughnuts, fruit pies, and many kinds of breakfast pastries. For richer products, such as roll-in pastries, where butter is a traditional ingredient, margarines made without milk-based additives can often be substituted with satisfactory results. Imitation whipped creams without dairy ingredients are available for use as adjuncts in the fancier kinds of sweet goods. Cream pies (banana, chocolate, coconut, etc.) could be a problem, since the traditional custard base relies heavily on the physical and organoleptic properties of milk and cream — use of a starch-based pudding mixture should allow the preparation of a lactose-free filling, but a close duplication of the traditional texture and taste will probably be difficult.

High-quality confectionery, such as chocolates with buttercream fillings may require major reformulation. Milk chocolate, which by regulation must contain substantial amounts of milk solids, could perhaps be reformulated with reduced lactose milk, but the results are hard to predict. Of course, the amount of lactose consumed in a day as a result of eating milk chocolate must be very small, except for real chocoholics, and they probably have other problems to take their mind off any minor discomfort due to lactose in the lower intestine.

Coeliac-disease

Celiac (coeliac) disease, which has also been called gluten intolerance and non-tropical sprue, is a disorder related to ingestion of materials containing the proteins (actually, gliadin) of wheat and some other cereals. It can be triggered by flours or other edible products from wheat, rye, triticale, durum, or oats. Some writers have said that barley or even malt also cause the typical symptoms of this disease, but there does not appear to be a great deal of evidence for this claim. Corn and rice products are probably low-risk foods for sufferers from celiac disease. It is important to recognize that cereal technologists and the medical profession differ in their definition of gluten; for persons dealing with cereals as ingredients in foods and feeds, gluten generally means the complex of proteins that give the elastic, cohesive, film-forming, and water-absorbing properties characteristic of wheat flour, while medical doctors consider gluten to be the prolamin from any grain that can cause celiac disease. It should be mentioned that "prolamin" is a definition based primarily on the solubility characteristics of a protein (or group of proteins) that may vary substantially in structural and chemical features depending on the source from which it has been obtained.

Etiology, symptoms, and distribution — This disorder is not an allergy; there have been reports of gluten allergy but it is less common than celiac disease and the symptoms are quite different. The disease has serious consequences, and is evidently rooted in a genetic predisposition, the symptoms generally first becoming evident in infants who are just beginning to eat solid food. A precise and incontrovertible demonstration of the mechanism by which gluten exerts its noxious effects has not been given. Troncone and Auricchio (1992) reviewed the theories found in the literature. They seem to favor the immunologic theory, which would lead to the viewing of celiac disease as a sort of allergy, though not a typical one.

A conclusive diagnosis of celiac disease can be made by finding flattened villi in the small intestine of persons who have been consuming gluten-containing foods. The symptoms of the illness follow from this damage to the mucosal lining and most of them result from failure to absorb nutrients. In children, the symptoms typically include abdominal distension, vomiting, diarrhea, wasting of muscle tissue, and failure to grow and develop at a normal rate. In adults, diarrhea is the main symptom, but fatigue, weight loss, anemia, cramps, and swelling of the tongue may be observed. Eventually, neurological disturbances, bone wasting, infertility, an increased risk of cancer can occur. Adults exhibiting symptoms of the disease have probably been susceptible since birth (Skerritt et al. 1990). Heiding et al. (1995) reported celiac disease first becoming apparent in a 47-year old man without previously known illness. The symptom initially presented was acute bleeding in the legs, with no known cause. Biopsies and other tests showed celiac disease. The patient quickly improved after he was put on a gluten-free diet and given supplements of various vitamins. The authors concluded, "Celiac disease can take a clinically unremarkable course for a long time and may finally become manifest through an isolated abnormality, such as bleeding." Troncone and Auricchio (1991) say that, in addition to the gastrointestinal symptoms, liver, joint, hematological, gynecological, and neurological symptoms are increasingly being recognized, and that a number of diseases have been found to be associated with gluten-sensitive enteropathy — among them are IgA deficiency, IgA nephropathy, sarcoidosis, insulin-dependent diabetes mellitus, as well as other autoimmune diseases. They also say that a greater than hitherto expected number of individuals have a clinically silent disease, and many others have a preceliac condition.

The incidence of gluten intolerance apparently differs among ethnic groups, ranging from high levels of one in 200 to 500 people in parts of Ireland, Austria, and Scandinavia to perhaps one in 1,000 to 5,000 people in other parts of Europe and in North America. The reported incidence would obviously be much less in countries where medical treatment is less common and opportunities for diagnosing the disease less frequent. Also, in many parts of the world, wheat is not a major food — rice, millet, sorghum, corn, or non-cereals (e.g., potatoes, taro) being used as the main carbohydrate sources, so the disease is never observed.

Gluten-free foods — There must be a very limited market for prepared foods meeting the dietary requirements of gluten-sensitive persons. Distribution of such specialties through super markets is probably highly impractical. Even in the best circumstances, the volume of sales would be so small as to make the turnover inadequate to support distribution in fresh form. Supermarkets would never allocate shelf-space to products having such a low turnover rate and the short shelf-life of fresh baked products. Frozen, canned, irradiated, or other preserved foods in ready-to-eat form, or dry mixes for consumer preparation would seem to be indicated regardless of whether supermarket, specialty store, or mail order sales are contemplated.

Removing gluten-containing materials from formulas for baked products such as bread, rolls, cakes, biscuits, lead to many seemingly unsolvable problems. It is futile to attempt to make a normal-appearing loaf of bread that contains no gluten. The typical form, structure, and texture of breads are based on gluten networks (or foams) expressed on the macro scale as assemblages of elastic bubbles inflated by gas. No other edible substance functions in exactly the same way as gluten. Of course, vital wheat gluten can't be used either, since this material contains most of the proteins of wheat flour, and it is protein that causes the disorder.

In one of the earliest publications describing tests of gluten-free doughs formulated to give simulated bread loaves, Jongh et al. (1968) described a series of experiments in which the structure-forming properties of wheat starch and cassava starch in a fermented dough were examined. They obtained their best results when 5% egg albumen was added to the dough; monoglycerides and other emulsifiers, as well as triglycerides (in this case, lard) improved results. No quantititative data on loaf volumes and crumb texture were given, but the photos seem to show significant differences from normal wheat flour loaves, even in the best examples.

The best non-wheat substitutes for bread and rolls (and they are not really satisfactory to most consumers) are based on eggs as the structure-forming ingredient and rice flour or corn starch as the cereal or starch component. Sometimes potato starch or the like is used as the bulking agent in order to eliminate cereal protein that might be present in trace amounts in commercial corn starch, but, usually corn starch or even corn flour forms the major constituent of most gluten-free breads. At best, such loaf breads resemble pound cakes or sponge cakes more than they do yeast-leavened wheat breads.

Most of the egg-based formulas will not give a batter that can be relied upon to support itself through the baking process when it is in the form of a one-pound loaf. Individual dinner rolls or muffins are more practical. Generally, the smaller the piece size, the better the appearance. Normal fermentation and proofing steps are out of the question for such products, although the addition of a fermented liquid at the final mixing phase might be helpful. In small-scale production, de-gassed beer (preferably a low-hops content beer), added as a minor percentage of the total liquid, might simulate some of the notes of normal bread's flavor, but the possibility of wheat protein being present in the beer must be considered.

A company on the west coast, Ener-G Foods, is making wheat-free loaves based (apparently) on tapioca starch. These are promoted as having a 12-month shelf-life as a result of being packed in hermetically sealed-pouches made of a silica-coated film that excludes oxygen. Another variety is described as being "yeast-free" as well as wheat/gluten-free (Reynolds 1996).

The following recipe for a gluten-free muffin can serve as a starting point for experimentation, but the author has not verified its performance.

Mix 9 parts water and 18 parts finely granulated rice flour with 2.5 parts compressed yeast. Bring the dough out of the mixer at as near 78°F as possible and allow to ferment for two hours. Add 27 parts water and 36 parts rice flour to the "sponge" and mix until smooth. Ferment for 45 to 60 minutes. Scale into muffin tins dusted with corn starch, filling cups about half full. Proof for about ten minutes. Bake at 385° to 410°F. This batter can't be processed on bread equipment.

Rusks, toasts, and flatbread of the Ry-Krisp type might be susceptible to re-design in a non-wheat format. Non-sweet biscuits (crackers) should be possible. An acceptable substitute for masa tortillas could probably be made using a starchy component, perhaps rice flour, with various texture modifiers — the need for such a product would be based on the assumption that the corn ingredient triggers symptoms when consumed by celiac sufferers, which may not be the case. Of course, wheat flour (also called white flour) tortillas would be unacceptable to susceptible persons.

ALLERGIES

This seems to be the age of allergies. One hardly qualifies as being politically correct unless he or she has two or three allergies. In the "olden days," we blamed goldenrod and other weeds, pollen from which was the cause of sneezing, coughing, weeping attacks in the late summer — in short, hayfever. Things have become more complicated. Now, it is a rare food or ingredient that has not been indicted by the The National Enquirer or other peer-reviewed journals as a cause of allergic attacks. In a recent autobiographical novel (no, that's not a mistake), a relatively famous ex-director of films explained that one of her husbands couldn't eat in certain restaurants because he was allergic to pectin.

True Allergies Related to Contact with Foods

According to specialists in the field, allergic reactions have a fairly well defined mechanism involving participation of the immune system, and food allergies can almost always be traced to the ingestion of a specific protein (and, perhaps, rarely, peptides and amino acids). Experts in the field estimate that only about 2% of adults, and perhaps 2% to 8% of children, are truly allergic to any food, and the majority of people with true food allergies are allergic to less than four foods (Hingley 1993). Some authorities place the actual figure at less than 1% of the total population. There is a much larger proportion of the population that suffers occasionally from reactions to relatively common foods. In only a few cases will the formulator and processor of foods for the general consumer market need to be concerned about these problems. Some examples will be described in the following section.

It is well accepted in the medical profession than infants can outgrow some allergies. One study (reviewed by Sampson 1992) indicated that 85% of infants outgrew their food sensitivity by their third birthday. Older children with allergies may not lose their hypersensitivity and it is not clear that adults lose their allergies, at least, not without treatment. Patients with allergy to peanuts, tree nuts, fish, and shellfish generally have very long-lasting hypersensivity.

Chemistry of Food Allergies

Most of the discussion under this heading has been adapted from the article by Taylor (1992).

Food allergens have been defined as substances that initiate immunological reactions recognized as food allergies. The primary immunological mechanism is probably a reaction mediated by a specific immunoglobulin E, designated IgE, a blood constituent. The sensitization process involves an allergen stimulation of the specific IgE. Upon a subsequent contact of allergens with IgE on the surface of sensitized mast cells and basophils, histamines and several dozen other mediators of the allergic reaction are released by the cells.

Proteins are virtually the only chemical compounds that act as allergens. Fats and carbohydrates rarely, if ever, cause allergic reactions, although it would be expected that commercially available peanut oil (for example) might contain traces of protein that could cause allergic reactions in susceptible individuals. One RNA, from shrimp, is believed to act as an allergen. Pure inorganic substances never react in this way, and, probably, individual amino acids never act as allergens since this would result in an situation incompatible with normal life processes in all persons; it is not entirely clear whether or not some peptides cause allergic reactions.

According to Taylor, small proteins are less immunogenic, thus have less allergenic potential, than large proteins. He says a molecular weight of 10,000 may represent the lower limit for an immunogenic response. Upper limits may be dictated by intestinal permeability. "Proteins larger than a molecular weight of 70,000 are less likely to be efficiently absorbed through the intestinal mucosa to obtain access to the IgE-producing cells . . . " Food allergens are resistant to many of the usual forms of food processing, including heat treatment, and (at least in some cases) survive conditions that cause proteolysis and denaturation of many proteins (Wahn et al. 1992). Deadly anaphylactic reactions may occur as the result of the ingestion or inhaling of minute amounts of allergens.

Most of the enormous number of proteins in foods do not trigger IgE reactions; allergens are found predominantly in foods of plant and marine origin, proteins of animal origin (with the exceptions of egg and milk), are rarely implicated in food allergies.

Symptoms of Allergic Reactions

An allergic reaction results from the ingestion of a substance, such as peanut butter, containing a chemical compound that causes the body's immunological system to mediate a hypersensitive response resulting in illness or even death in the susceptible individual. The reason why some people become ill after eating a food that most other persons find to be completely innocuous is still not known. One of the best short reviews of the sequence of events in the body's reaction to allergens is that of Taylor (1992), which should be consulted if the reader desires further information on this subject.

The sites most frequently exhibiting symptoms of allergic reaction are:
mouth — swelling of the lips or tongue; itching lips.
airways — wheezing or breathing problems.
digestive tract — stomach cramps, vomiting, diarrhea.
skin — hives, rashes, or eczema.

Even the slightest reaction is likely to be quite unpleasant, but the greatest danger comes from anaphylaxis (also called anaphylactic shock), a violent generalized reaction affecting several parts of the body simultaneously. Common symptoms include difficulty in breathing, a feeling of impending doom, swelling of the mouth and throat, drop in blood pressure, and loss of consciousness. Death may supervene in a short time. Very small amounts of a substance to which the person has become sensitized can cause anaphylaxis. It can occur in as short a time as 5 to 15 minutes after ingesting the allergen.

Common Foods that Cause Allergic Reactions

Common foodstuffs known to cause allergic reactions in some individuals include: wheat, rice, milk, eggs, nuts (including peanuts), soybeans, buckwheat, and many kinds of marine foods (e.g., shrimp, clams, and fish). Other, less frequently reported examples are fruits such as peaches and papaya, and vegetables such as green peas and potatoes. Meats, i.e., muscle tissue from land animals including fowls, are seldom implicated in food allergies. A brief list of foods known to cause these symptoms follows:
Legumes — especially peanuts and soybeans.
Crustacea — shrimp, crabs, lobster, crayfish.
Milk — including cows' milk and goats' milk.
Eggs — from almost all avians.
Tree nuts — Almonds, walnuts, Brazil nuts, hazelnuts, etc.
Fish — cod, haddock, salmon, trout, etc.

Molluscs — clams, oysters, scallops, etc.
Wheat

Milk is the most common vector of allergic reactions in young children. Some studies have shown that slightly more than two percent of infants are allergic to cows' milk. Goat's milk and probably milk from some other species can also cause these symptoms. Eggs are also important sources of allergens, both the white and the yolk containing proteins that act in this way.

The major allergenic food affecting adults is peanuts, and this legume is important in children's diseases as well. Soybeans also contain multiple allergens, but are less important overall in human illnesses, possibly because the protein fraction of soybeans is not a common food for humans (being mainly used in cattle feed); its use in foods is increasing, however. Of course, not all allergic reactions result from foods, but only the latter will be considered in this book.

Responsibility of the Manufacturer and Packager

Surely not even the most paranoid consumer advocate would expect food producers to remove every possible allergen from their products. This would require removing essentially all proteins, peptides, and amino acids, since it is at least a possibility that any one of these could cause allergic reactions in a person who has, for some reason, become sensitized to one of the nitrogenous compounds or a breakdown product of it. This would leave us with the option of trying to subsist on chemically pure sugar and salt, and a few purified fats. It is reasonable to expect, however, that persons who are allergic to some substance would have the opportunity to protect themselves from ingestion of this material by reading the ingredient declaration.

The manufacturer has a responsibility to declare on the label in clear terms the ingredients used in the formula of the product. There is little question about this, the law requires it, but problems do arise as a result of inadvertent contamination of the product by allergens that are not intended to be in the finished product and so are not declared on the label. It may also be misleading to include non-specific terms on the label, such as flavors (both natural and artificial flavors can contain proteins).

Use of rework, scrap, formulation mistakes, stale returns, and food recovered from wrongly labeled packages may introduce allergenic materials that do not appear on the labels. Food preparation methods, including but not limited to the use of shared weighing, mixing, and cooking equipment, frying oils, and baking pans can also inadvertently contaminate batches with constituents not appearing on the labels of finished products. According to Taylor and Nordlee (1996) several deaths have occurred from such practices in the food service industry. These authors give the example of processing egg-free pasta and egg noodles on the same machinery, presumably making the change without adequate cleaning of the press.

Contamination with an allergenic substance often goes unrecognized unless the contaminated product is consumed by a susceptible individual, and then is recognized only if the victim is suspicious that such contamination might have occurred and is sufficiently motivated to make an investigation.

According to Pape (1996), "One idea worth considering is prophylactic labeling, that is, the inclusion in the ingredient listing of ingredients that may be present in the food unavoidably (for example, 'may contain peanuts'). Consumers with known allergies are usually dedicated readers of food labels and would be alerted to the potential risks. Food companies would avoid the need to recall products which are safe for the vast majority of consumers but which pose risks for the population allergic to a specific unlisted and unintended ingredient."

Workers in food industries may be exposed to substances that are allergenic. Sensitizing agents include flour, green coffee beans, and shellfish. Exposure can occur by inhalation and/or contact. Diseases associated with workplace exposures are occupational asthma, hypersensitivity pneumonitis, and dermatitis (Lehrer and O'Neil 1992).

chapter twelve

FOODS MEETING RELIGIOUS OR PHILOSOPHICAL TENETS

There are large numbers of consumers that base their purchase and consumption of at least some of their food on the product's conformity to their religious or philosophical beliefs. The degree to which the believers adhere to their principles varies, but a devotee of organic or natural foods will certainly give precedence in his choice pattern to a product which is alleged to consist only of organic or natural foods, even though his final decision also may be affected by economic and availability factors.

PHILOSOPHICAL RESTRICTIONS

Vegetarians and persons who try to restrict their diet "natural" or "organic" foods form two separate, but often overlapping, segments of the customer population that are worthy of consideration by the food product developer. In many cases, they are willing to pay substantial premiums for foods meeting their requirements, an inducement for entrepreneurs who would otherwise be reluctant to spend time and money trying to break into this limited market.

Vegetarians

McCue (1996) quotes a 1994 Roper poll as showing that about 0.3% to 1% of the U.S. population chooses to never eat meat, fish, or fowl. However, 3% to 7% of the population consider themselves to be vegetarians.

A wide variety of vegetarian foods are available. One company offers complete dry packaged meals that can be prepared by adding boiling water and simmering. meals Sweet 'n Sour Dinner, Mexican Style Dinner, Italian Style Dinner, Stroganoff Style Dinner, Pizza Style Dinner, A La King Dinner, Western Chili Dinner, and Souper Dinner. Ingredients for the Sweet 'n Sour Dinner are enriched rice, textured soy protein concentrate, sugar, natural and artifical flavors, modified corn starch, onion flakes, dried bell peppers, lactic acid, white pepper, garlic powder, disodium inosinate, and disodium guanylate.

Imitation meat – Most imitation meats used in vegetarian entrees are based on soy protein concentrates that have been formed into textured pieces by extrusion under high pressure and elevated temperature. This process aligns the fibers so that a chewy structure reminiscent (to some consumers) of muscle tissue results. In addition to use of such materials in totally vegetable meat imitations, they are also mixed with inexpensive ground meat to act as extenders, cheapeners, and diluents of the cholesterol and saturated fats in beef and pork.

Vegetable protein may be used at levels of up to 30% of certain meat products offered in the Child Nutrition Program of the federal government. It is also found in some pizza toppings as meat extenders, usually in combination with sausage bits or crumbled hamburger. Soy protein derivatives are also fairly common in Mexican foods, e.g., in some brands of beef and bean burritos and in tamales.

Suppliers can furnish flavors that do a remarkably good job of imitating the taste of almost every kind of flesh foods cooked in a variety of ways — roasted, fried, boiled, or grilled.

Organic and Natural Foods

It is reasonable and appropriate to consider organic and/or natural foods in this chapter, since the only justification for creating a special legal category for these foods is a philosophical, in some cases apparently religious, concept, namely that "nature" somehow is more benevolent, safer, better adapted to human needs, indeed more caring, than science, food producers, or food sellers.

The terms are not precise. From a scientific point of view they are nonsensical, but from a marketing viewpoint they it make a lot of sense. According to Allen (1995B), the organic food category accounted for 5.5% of all new items introduced during the first half of 1994. Sales have grown from about $178 million in 1980 to about $1.89 billion in 1993. These figures include prepared foods, produce, and bulk ingredients.

An "organic" food is generally thought to mean a product that has been grown in an environment managed and isolated so that it is free of pesticides, herbicides, hormones, and other manufactured chemicals. Some adherents to this belief regard natural pesticides (pyrethrum, for example) as being all right, while others are not so sure. The Organic Foods Production Act (OFPA) was passed by Congress in 1990; a compliance date of October 1, 1993 was established but it was not met. A proposed rule is expected to be offered sometime in 1996.

The basic rules established by the OFPA are:

(A) General requirements. Food products labeled as "organic" must not contain any synthetic ingredients or processing aids unless they are specifically allowed. Natural substances found to be toxic will be prohibited. Approved synthetic substances and prohibited natural substances will be included on the National List. Ingredients such as fruits, vegetables, and grains must be obtained from certified organic growers. A three-year transition period between conventional and organic growing methods will become standard.

(B) Exceptions. Products labeled "organic" must contain 95% organic ingredients by weight. Products with greater than 50% but less than 95% organic content will be able to carry restricted organic labeling. Water and salt do not have to be derived from organic sources.

(C) Prohibited substances. Added sulfites, nitrites, and nitrates will be prohibited, as will ingredients containing excessive levels of nitrates, heavy metals, or toxic residues, and any ingredients derived or produced through biotechnology.

(D) Special rules apply to livestock, poultry, and dairy products. Livestock and poultry to be slaughtered must be raised on organically produced feed. Growth hormones, sub-therapeutic does of antibiotics, and synthetic internal parasiticides are to be prohibited. Milk and milk products having organic labeling must be taken from animals raised and handled under the same requirements as for livestock, for at least twelve months.

(E) Exemptions. Retailers that do not process or manufacture any of their own products, and farmers that earn less than $5,000 per year are exempt from certification. Even so, they must follow all of the standards (Allen 1995B).

The OFPA directed the USDA to establish a national organic certification program that would certify growers and processors who wanted to sell "organic" foods. State and independent certifying agents would be accredited by the USDA. At the present ime, 28 states have passed laws relating to organic foods. Sixteen of these prohibit organic labeling without certification, ten states have their own certification prgrams, and six use independent certification agencies.

There are many unknowns and conflicting views relative to the exact requirements. Questions relating to the acceptability of synthetic vitamins have not been answered to the satisfaction of all, but it appears that these compounds are acceptable, especially if there are laws on the books mandating the enrichment of certain foods and a synthetic material is the only practical alternative.

RELIGIOUS RESTRICTIONS

Many of the world's religions establish restrictions on the foods and beverages to be consumed by their adherents, and, often, on the cooking and serving procedures. Limitations of space allow us to include only a few of the most important (from a marketing standpoint) dietary systems in this survey. Only those aspects of religions affecting the manufacturing and marketing of foods within the US will be included in the following discussion.

Kosher

Jewish dietary laws divide foods into four categories or "genders." The first three categories are kosher foods, while the fourth includes all foods that are not kosher.

The first category is "meat" from ruminants and most domestic birds, provided the animals have been slaughtered according to religious precepts. The slaughter process and further treatment (salting and soaking) is intended to remove the blood, which is a prohibited substance. products derived from meat remain meat, and in addition all processing equipment that comes into contact with the meat becomes "meat."

The second category is "dairy." Milk obtained from kosher animals, all processed products from milk, and all equipment that comes into contact with milk become "dairy."

The third category is "parve," also called "pareve" and "neutral." Food or equipment that is kosher but neither meat nor dairy is parve. This includes all plant materials, eggs without blood spots, and fish with fins and removable scales. Grape products fall under a separate set of rules, effective for example in wineries.

Dairy and meat products cannot be mixed, either in processing, home preparation, serving, or in the stomach.

For Passover, any material made from wheat, oats, rye, barley, and spelt are prohibited, except for the particular type of unleavened bread made specifically for the holiday (matzoth). A number of materials including corn, rice, mustard, buckwheat, and legumes, have also been prohibited for Passover by the rabbis (Chaudry and Regenstein 1994).

Modifying non-kosher products so that they conform to Jewish dietary laws can lead to substantial ingredient changes, processing problems, and scheduling difficulties, particularly if the manufacturer also intends to make non-kosher products with the same equipment or even in the same plant. The kosher producer will almost always want to place on the product label a guaranty mark of some kosher-certifying person, agency, or group, and in such cases the factory management will have to accept inspections of the plant, ingredients, ingredient specifications, products, packaging, cleaning materials, etc., by representatives of the agency. And, you will not be able to keep your formulas confidential.

The manufacturer will also have to pay a fee to the certifying group, generally a fairly small amount if the per-package levy is considered but the total amount required can be substantial. Requirements of different certifying organizations vary, and they can become quite complex. It is advisable to get full details of their requirements from the specific agency whose kosher-guaranty mark you intend to use before any long-term contractual commitment is made.

The Orthodox Union of New York is one of these businesses. As a first step they will check a list of a food's ingredients against their computerized list of materials. They classify ingredients such as cane sugar and corn starch as Group I, i.e., substances not requiring letters of certification from the supplier. For more questionable ingredients, such as shortening, the Orthodox Union will provide the manufacturer with a list of suppliers of kosher shortenings. Clients will be subjected to periodical visits by an inspecting rabbi (mashgiah), or, in some cases, to continuous on-line supervision (LaBell 1995).

As with any doctrinal system that has been in existence for a considerable time, modifications, clarifications, extensions, and differences of opinion have arisen among the purveyors of kosher certification. For example, there are different degrees or types of kosher. Pareve means, generally, the product contains neither dairy nor meat products, for example canned peaches, granulated sugar, or frozen broccoli prepared under kosher conditions. If the article is claimed to be kosher-for-passover it must be completely free of leaveners, as defined by the authorities. "Glatt kosher" meat, in addition to being processed under the usual slaughter-house supervision by rabbis, must be essentially free of lung adhesions. The processing of "Cholov Yisroel" milk is supervised throughout the entire system by kosher overseers.

Kosherizing authorities sometimes disagree. Examples include (1) so-called "kosher gelatin" is not considered kosher by certain kosher supervisory agencies, (2) the kosher status of vodka was the subject of a dispute between the only two rabbis in Poland (Newman 1993). The latter disagreement was resolved by some of the 25 or so vodka producers in Poland getting their bottles stamped by non-Polish rabbis. The director of regulatory affairs for a supplier of 400 kosher seasonings and sweetenings said (quoted by O'Donnell 1993), "Our biggest frustrations are the lack of acceptance of kosher certifications between rabbinical groups and an understanding of what we must do. For example, our supervisory service [rabbinical certifier] agrees to accept the certification of an ingredient by another group. We then purchase that ingredient and use it in many of our formulas. Later, we want a new product certified but the original ingredient from the same company is no longer accepted as kosher." The answer is, of course, that these kosherizing individuals and groups are competitive businesses and few companies are interested in advancing the interests of their competitors.

If your product purports to be kosher, expect the government to become more interested in your business. According to Regenstein and Regenstein (1992), "Seventeen states currently regulate kosher foods, specifically referring to Orthodox standards in one way or another." The rather obvious breaches of church-state separation that such regulations seem to indicate are addressed by the Regensteins at considerable length in the quoted article.

It is not sufficient that the ingredients and processing method be kosher; the packaging material containing the food must also conform to the rigid dictates of Mosaic/Talmudic law as interpreted by the rabbis. Presumably, petroleum derivatives are basically kosher, but additives and processing aids used in making resins and forming plastic packages may contaminate the container and the foods in them with unpermitted substances. If a plastic is prepared and formed using, for example, tallow-based additives (e.g., calcium stearate, zinc stearate, and glycerol monostearate), it may be considered that these materials can migrate to the surface of the film or container and thus come in contact with the food. A possible solution is to replace tallow-derived stearates with similar additives made from vegetable oils (Pszczola 1994 and Ainsworth 1994).

Kosher means ceremonially or doctrinally "clean," i.e., conforming completely to a set of religious requirements. It has nothing to do with sanitation. Recent attempts to position kosher foods to the general public as somehow more wholesome or sanitary than non-kosher foods (Hwang 1993) have no rational foundation. Kosher foods are almost always more expensive, sometimes much more expensive, than comparable foods not conforming to kashrut. Art Berman, president of Rokeach Food Corporation, a long-time leader in the marketing of kosher foods, says (quoted by Otto 1993), "The traditional kosher market has not really grown over the last few years. It has been stable."

Islamic Food Laws

The approximate Islamic equivalent of the Jewish "kosher" is "halal," while "haram" means unlawful or prohibited. There are several details that are almost the same in the Jewish and Islamic systems, while in other respects they are very different. Although pork products and blood are forbidden foods in both cases, alcoholic beverages are prohibited in the Islamic system but can be kosher, etc. (Chaudry 1993). More details can be found in *Islamic Dietary Laws and Practices* (Hussaini and Sakr 1983) and *The Lawful and the Prohibited in Islam* (Al-Qaradawi 1984).

There is at least one organization (Islamic Food and Nutrition Council of North America) providing certification for halal foods, and it gives the opportunity for applying a trademark or "seal," which is in appearance basically a crescent moon inclosing the letter M, to packages of food. The following discussion is basically a paraphrase of the writings of Dr. Muhammad Chaudry, president of the IFNCNA (Chaudry and Regenstein 1994, Chaudry 1992).

All things are made halal by Allah, except those things that have been specifically prohibited in the scriptures or through extrapolation of the direct guidelines. The meat of ruminants and other herbivores is halal, provided the animal has been slaughtered humanely while pronouncing the name of Allah at the time of slaughter. The meat of chickens, turkeys, and many other domesticated and game birds is halal when such birds are slaughtered ritually or killed according to acceptable procedures. Most fish and certain types of seafood are halal without any ritual slaughtering, if the killing is done humanely. Dairy products and eggs from permitted animals are also halal. All foods derived from plants are halal, unless they are poisonous or intoxicating. Foods derived from the mineral kingdom (e.g., salt, sodium bicarbonate, petroleum derivatives) are halal except those that might produce intoxication or which pose health problems.

There are no restrictions on combining foods from different halal groups. Islamic food laws emphasize cleanliness, sanitation, and purity. Food preparation equipment, utensils, and packaging materials must be clean and free from prohibited substances and contaminants.

Eating haram foods is absolutely forbidden for every Muslim except under extremely exceptional circumstances. Haram things can be grouped in the following categories:

1. Carrion or meat of dead animals.
2. Swine and their by-products.
3. Blood, whether flowing or congealed.
4. Anything immolated unto idols or dedicated to anyone other than Allah.
5. Intoxicants, such as alcohol and drugs.
6. Carnivorous animals and birds of prey.
7. Land animals without ears, such as snakes and frogs.
8. Animals that have been killed in a manner that prevents the blood from being fully drained from the carcase.

By deduction and by extension from the above categories, certain other animals and products are also considered haram.

"Mashbooh" (also, "mash-booh") means "suspected" in Arabic. If the origin of a food item is in doubt, or there is uncertainty about its conformity to Islamic laws, the food is mashbooh — a grey area between what is permitted and what is prohibited. Many products in the marketplace fall into this category. Muslims avoid such products until they make certain through investigation that a certain product is indeed halal.

Production of food products by fermentation processes, e.g., the manufacture of cheese, bread, fermented milk, vinegar, and the like is not prohibited. It is the use of the products so produced that is permitted or prohibited according to the scriptures. The use of alcohol as a substrate, intermediate compound, solvent, or precipitating agent may be permitted as long as the level of alcohol remaining in the material that becomes a food ingredient is present at the minimum level technologically possible. If purified food chemicals produced through biotechnological techniques have traditional equivalents that are halal, they are also halal; for example, monsodium glutamate, citric acid, and lactic acid.

Enzymes may or not be acceptable depending on the source. Bovine rennet collected from calves that have not been slaughtered according to Muslim requirements is not accepted by most Muslims, but chymosin (the main enzyme found in rennet) that has been produced microbiologically through transcription from the bovine chymosin gene is universally accepted.

Hindu Food Preferences

Hinduism is a "synthesism of diverse and competing faiths" (Kilara and Iya 1992). It is not organized into a monolithic regulatory structure. As a result, the dietary habits and preferences of Hindus vary greatly from caste to caste and for other reasons. In general, there are strict rules governing the preparation of foods, sharing of foods, consumption of foods, etc. Brahmins (the highest caste) are almost always vegetarians, at least ostensibly.

Christian Denominations

Many Seventh-day Adventists are vegetarians, or have inclinations in that direction. It is not difficult to meet their requirements with most bakery products, the avoidance of animal shortenings such as lard or ingredients derived from beef fat would be among the major consideration.

Although some members of the Church of Jesus Christ of Latter-day Saints are vegetarians, total avoidance of meat products is not, in fact, a tenet of this religion. Moderation in consumption of meats is counseled, however. This should not create a problem in preparing bakery foods for this market. Avoidance of alcoholic beverages, coffee, and tea is obligatory. Note that chocolate products are not prohibited, though some members do avoid them through a mistaken impression that caffeine and like stimulants are prohibited. Making bakery products without alcoholic beverages, coffee, or tea is not a problem. These ingredients seldom appear in formulas, though they are found in a few gourmet products such mocha icings and beer bread.

chapter thirteen

FUNCTIONAL FOODS AND SUPPLEMENTS

Recently, there has developed a fad of calling foods designed to have some sort of beneficial physiological effect, "performance" or "functional" foods. Other terms that have been used are "designer foods," "nutriceuticals," and "pharmafoods." It is not clear whether the different terms are intended to have nuances not immediately apparent or are merely synonyms for diffuse and poorly understood categories that overlap, sometimes being entirely congruent. Peter Barton Hutt, formerly of the FDA, points out that there is no regulatory category of "functional foods" or "nutriceuticals."

Related to nutraceuticals, and in some respects indistinguishable from them, is the group of substances and preparations called "supplements." The principal difference is that supplements would not ordinarily be called "foods" because they serve no recognized nutritional purpose or, if they do, it is an inconsequential part of their appeal — they would never be considered part of a meal or even as a snack. . They resemble pharmaceuticals more than they do foods, though they are not regulated as pharmaceuticals.

FUNCTIONAL FOODS

According to Hasler (1996), "The past decade has seen an explosion of interest in foods for the prevention of disease or the promotion of optimum health, also known as 'functional foods.' In order for this field to successfully evolve, however, there are a number of scientific regulatory and business hurdles that need to be overcome. First, a concise definition for this new and emerging product category is essential. Second, how these foods will be regulated needs clarification, particularly in light of the Nutrition Labeling and Education Act of 1994 (NLEA) and the Dietary Supplement Health and Education Act of 1994 (DSHEA). Currently, there is an unlevel playing field between dietary supplements and conventional foods. Finally, there is a lack of appropriate reward (i.e., the market exclusivity) for industries investing in the research and development of functional foods . . . "

It is the purpose of this section to review the current status of functional food marketing and to suggest ways of approaching the formulating and processing of products of this type.

Products in these categories have been defined as "foods which affect physiological functions of the body in a targeted way so as to have positive physiological effects because they contain ingredients that may, in due course, justify health claims" (Roberfroid 1995). The term,"functional claims," refers to the effect of a food or ingredient on a specific physiological/biochemical process with no fully understood relationship to health. An example that has been given is, increasing the number of bifidobacteria in colonic flora, which is a functional effect that has not been conclusively shown to improve health.

According to the cited author, a food can be made functional by using one or more of the following approaches: (1) eliminate a component having negative effects such as activating allergies or causing mutations, (2) increase the concentration of a component with a beneficial effect, e.g., antioxidant vitamins, essential fatty acids, minerals, or dietary fiber, or (3) add a component with beneficial effects, such as probiotics, non-digestible oligosaccharides, and antioxidants of vegetable origin, and (4) partly replace a food ingredient which, in excess, is likely to have a negative effect on physiology. Presumably, also included would be foods designed to promote weight gain, foods advertised as being muscle-builders or general enhancers of athletic ability or sexual performance, and isotonic beverages claimed to have special advantages in maintaining the blood's ionic status at a desirable level.

The Functional Foods for Health program at the University of Illinois maintains an online database, described as a "natural product alert," containing well over 100,000 publications selected from those thought to be helpful to medical and pharmaceutical researchers. Collaborating in this effort are the U. of Illinois School of Medicine in Chicago, the University's Food Science and Nutrition departments, and several representatives from the food industry. Dr. C. M. Hasler is listed as the Program Director.

NUTRACEUTICALS

Nutraceuticals (sometimes spelled "nutriceuticals") are foods (including ingredients) that confer medical or health benefits, including the prevention, treatment, or cure of diseases. The term is not precise, and it is sometimes extended to cover meal-in-a-can or meal-in-a-bar, which is as far as science has been able to approach the nutritional Shangri-La of meal-in-a-pill.

A medical food can make claims for efficacy without pre-clearance from the FDA. These claims can be outside the NLEA claims approved by the FDA, provided there is adequate documentation supporting the claims, and as long as the marketing is restricted in certain ways. A "dietary supplement" can claim to have an effect on body structures and functions, but can't claim to cure or prevent diseases. Obviously, there is room for varying interpretations of what these distinctions really mean — perhaps that was the intention of the regulators.

The Dietary Supplement Health and Education Act (DSHEA), which was passed in 1994, was meant to prevent the FDA from taking control of vitamin and mineral supplements, and from controlling what the agency had said it regarded as excessive fortification and supplementation with individual nutrients. In essence, the Act restricted the FDA's power to classify dietary supplements as drugs, unless the agency could prove that specific combinations of vitamins, minerals, other micronutrients or alleged micronutrients, and herbs were dangerous it could not restrict their sale.

State agencies may have wider powers than the FDA in regulating some of the truly dangerous herbal mixtures that have entered the marketplace. There are reports that supplements including a herb called ephedra have caused at least one death. Florida has already announced a ban on some of the herbal products, and other states are contemplating action (Katz 1996).

In 1992, the Office of Alternative Medicine (OAM) was estalished in the National Institutes of Health. This new agency was given the mission of promoting scientific research of alternative health practices. In 1996, the OAM funded at least ten projects underway at several research facilities throughout the country. It is expected that conclusive monographs will be published on the efficacy of herbal ingredients.

Quasi-pharmaceutical Components of Foods

Many marketers want to get part of this action, so they seek out claims made for food components by herbalists, nutritionists, etc., then find products they can market that can take advantage of the claim. You find the compound, they'll find the need. Some of the claims that have been made for the health benefits of ordinary foods include:
Allylic sulfides from onions fight inflammatory activity.
Anticaries activity has been found in seeds of 11 species of plants.
Benzoic acid from cranberries alleviates urinary tract infections.
Beta carotene plays a role in preventing lung cancer.
Carotenoids in, e.g., carrots, may protect against cancer.
Cartilage from chickens reduces the symptoms of rheumatoid arthritis.
Catechins from green tea lower risk of gastrointestinal cancer.
Cumarins from parsley hinder blood clotting.
Daistein from soy is said to lower cholesterol.
Fiber from whole grains encourages growth of helpful bacteria.
Flavonoids from broccoli may block hormones that promote cancer.
Folic acid from leafy foods, etc., prevents neural tube defects in newborns.
Gamma-glutamyl alylic cyclenes from garlic raise immune system activity.
Glucomannan from konjac tubers increases glucose tolerance in diabetics.
Genistein from soy products deters proliferation of cancer cells.
Indoles from kale activate enzymes that inhibit estrogen.
Inulin from Jerusalem artichoke lowers cholesterol.

Isothiocyanates from horseradish induce protective enzymes.
Lactobacillus in yoghurt helps treat vaginal yeast infections.
Licorice contains substances that are thought to prevent disease.
Limonoids from citrus fruit activate protective enzymes.
Linolenic acid from flaxseed acts as an anti-inflammatory agent.
Lycopenes from tomatoes may prevent cancer.
Monoterpenes from cabbage inhibit cholesterol production.
Phenolic acids from berries may help the body fight cancer.
Phthalides from celery stimulate enzymes that detoxify carcinogens.
Plant sterols from eggplant block estrogen's role in breast cancer.
Polyacetylenes from parsley help regulate prostaglandins.
Saponins from beans may have anticarcinogenic properties.
Triterpenoids from soy products act against ulcer development.
Unidentified polymeric compound in cranberries and blueberries reduces adhesion of Escherichia coli to the urinary tract.

Discussions of a few of the compounds or materials derived from plants and animals on which particular emphasis has been placed in recent years will be found in the following paragraphs.

Coenzyme Q — A lipid that transfers electrons in the process of metabolizing food components, It has been the subject of several athletic performance studies, with no definite positive results.

Creatine — A nitrogenous substance, $C_4H_9N_3O$, found in the muscles of vertebrate animals, and also in the blood, brain, etc. It is made in the body and also absorbed from foods such as meat. Creatine has been added to various sports foods on the theory that it combines with phosphate to make energy available for short bursts of activity such as involved in sprinting or weight lifting.

Curcumin — Curcumin is a yellow crystalline substance that is the coloring principle of curcuma root, the latter being the source of the common spice curcumin, much used in flavoring and coloring mustard and some other condiments. Curcuma root has long been an item in the herbalists' and natural healers' armamentariums. It has received recent attention from Liu (1996) and others. Liu claims that curcumin is a potent drug for treatment and prevention of cardiovascular disease. Studies have demonstrated that curcumin reduces platelet-aggregation of blood of humans (69.0%) and blood of animals (20.9%); reduces cholesterol of serum (60.1%) and triglyceride of liver (40.2%) in high fat dietmice. Curcumin has demonstrated antioxidant activity including inhibition of xanthine oxidase (69.0%), and monoamine oxidase (64.5%), and radical scavenging activity

(75.0%). Meanwhile, curcumin can increase effects of lymphoblast transformation (81%) and phygocytosis of peritoneal macrophages (64.5%), and particle clearance by the reticuloendothelial system (89.0%) in immunosuppressed mice.

Gamma-aminobutyric acid — Gamma-aminobutyric acid, or Gaba as it has been called, is apparently effective in lowering blood pressure. The patent of Saikusa et al. (1996) describes Gaba-enriched food materials and methods for making them. The material was used to enrich green tea and several forms of rice germ.

Isoflavones of soybeans — Kuhn (1996) described the current status of research on phytoestrogens (plant hormones) found in, for example, soy protein. Researchers have been working on possible applications for the main soybean isoflavone, genistein, since it was shown in 1986 to inhibit cancer cell growth. Some workers think it might also lower blood cholesterol levels.

A former National Cancer Institute scientist who is now a consultant on soy use says that twenty epidemiological studies have suggested that as little as one serving per day of tofu contributes to a reduced cancer risk. Attempts are being made to show an improving effect of genistein on menopausal symptoms, on prostate cancer, on breast cancer, and on osteoporosis.

Lactoferrin — This is an antibacterial protein found in several animal secretions, as in milk and colostrum. The antibacterial effect is said to be due to the substance's chelating of iron and its effect on osmotic pressure of cells. Its high affinity for iron has led to the belief that it can help increase the oxygen capacity of the blood and provide increased endurance for athletes.

Although some lactoferrin is isolated from enzymatically treated cow's milk, most lactoferrin offered as an ingredient is not derived from animals; instead, it is harvested from aspergillus fungi that contain an implanted gene for the substance.

L-carnitine — Acetyl-L-carnitine has been described as an important dietary supplement by Lonza, a Swiss firm that synthesizes the material, and by other authorities. This amino acid compound is a derivative of L-carnitine, and both are found in human muscles, brain tissues, and hormone-producing tissues. Carnitine is a vital co-enzyme involved in fat metabolism; it is believed to carry fatty acids into the energy-producing compartments of cells. It has been suggested, in some articles going back 20 years, that carnitine derivatives stimulate the formation of neurotransmitter functions and may be helpful in stimulating cognitive function and improving the mental deterioration that often accompanies old age. It has also been claimed that carnitine spares glucose stored in muscle cells, prolonging endurance.

Melatonin — This substance has received more publicity and greater consumer acceptance than any other nostrum in recent years. Although it was originally promoted mostly on the basis of its sleep-inducing properties, it was only a matter of time before its mysterious properties were alleged to include other functions, such as delay or prevention of aging. Dr. William Regelson, Professor of Medicine at Medical College of Virginia, and author of the book, *Melatonin Miracle* (Regelson 1994) puts it this way: "Apart from clinical sleep induction, melatonin delays and reverses aspects of aging in mice and rats, possesses free radical scavenging activity, and may have value in the treatment and prevention of diseases, including cancer, epilepsy, colitis, and Alzheimer's."

It has not yet been recommended as a food supplement, but that awaits merely the passage of time.

Vitamins and Minerals from Natural Sources

Suppliers of ingredients rich in specific vitamins, usually vitamin C or vitamin A, claim advantages in labeling for foods containing their materials. Thus, it is believed a food having an ingredient label claiming acerola berry extract (very high in ascorbic acid) has more appeal to many consumers than an ingredient statement including "vitamin C." which some persons might regard as being synthetic (which the usual ingredient is, of course).

Replacement of synthetic vitamins (or vitamins produced in microbiological fermentations) by natural source vitamins almost always results in a substantial increase in ingredient cost. Also, the intrusion into the batch of foreign flavors (and sometimes colors) that are unavoidably present in the crude, or even partially purified, preparations containing natural vitamins, is seldom a positive feature.

Nonetheless, there are indications that natural sources of vitamin A may be superior in their health-giving qualities to the synthesized materials. In 1996, there was an announcement that beta-carotene and Vitamin A supplements did not prevent heart disease or lung cancer, and may have increased the cancer risk of persons who smoked. In another study, published about the same time, base on a panel of 22,071 male physicians, it was found tht beta-carotene supplements had no significant effect on incidence of cancer or cardiovascular disease. These results seem to confirm earlier results which showed an 18% increase in lung cancers and 8% more deaths in a group of male smokers taking a beta-carotene supplement for five to eight years.

One commentator on these rather rather surprising (for some) results, Susan Taylor Mayne, suggested that dietary (not from supplements) raw fruits and vegetables, and vitamin E supplements did reduce the risk of lung cancer in non-smoking men and women. As stated by Allen (1996A), "What science has yet to determine is whether the antioxidants are responsible for the protective effect or whether there are other factors involved. For example, people who eat a lot of fruits and vegetables may also get more exercise and eat less fat — both proven factors in a reduced risk of heart disease, and the subjects of a great deal of research on cancer risk."

Another point to consider is that the vitamin A obtained from natural sources consists of a number of active compounds and some precursors that may have different physiological effects than the more-or-less purified synthetic material available as a supplement. Additionally, the commercial product may, in fact, contain traces of impurities that are not encountered in nature, and which can have deleterious effects when consumed regularly over long periods of time. Whatever the reason, it is clear now, as it always was to many people, that consuming excessive quantities of these highly active substances is not a good idea.

Herbal Supplements

According to Toops (1996), the US herbal supplement market brought in $772 million in 1995. It is said that the European herbal trade amounts to about six billion dollars a year, with Germany and the UK accounting for over one-half of the total. Both these countries have systems for cataloging and testing herbals and regulating the industry within their borders through the European Scientific Cooperative for Phytotherapy, an EU committee. Toops (quoting Leiner Health Products literature) says 60% of consumers buy herbal supplements for "increasing energy," 56% for "preventing colds," 54% for "boosting immunity," and 43% for "improving sleep."

Katz (1996A) describes a melange called "Greens Mixture," marketed by a California company, that incudes powder from barley grass juice, brown rice powder, powder from wheat grass juice, spirulina, spinach powder, green tea extract, chlorella powder, echinacea extract, blue green algae, broccoli powder, and anti-caking agents. The blend offers a phenol activity of 18.0 from the green tea and 695 IU of vitamin A from the other ingredients.

Hazards accompany the use of new, or even old, botanicals in a foodstuff or dietary supplement. According to Dr. John Staba, Professor Emeritus of the Department of Pharmacognosy and Medicinal Chemistry at the University of Minnesota, the following questions should be asked:

(1) Is the plant quality within a minimal and maximal range of acceptability

from lot number to lot number?

(2) Do we know the botanical species used, the portion of the plant used, and its amount?

(3) Does it contain adulterants, such as toxic alkaloids?

(4) Does it contain microbial contaminants, such as toxic aflatoxins?

(5) Does it contain a high content of radioactivity, insecticides, or growth regulators because of the manner in which it was grown or collected?

(6) Can it be stored and processed without degradation — even high concentrations of water might cause the activeingredients to degrade.

(7) Can we provide a drug dose from a botanical so that clinical trials can be made and evaluations made meaningfully?

(8) Is it properly labeled so that the health professional, vendor, and consumer know what they are getting?

Aloe — Aloe vera is a spiny succulent, often mistaken for a cactus, that has been used for centuries for various medicinal purposes. Perhaps the most common use was as a source of a gel that could be applied to burns and abrasions as a soothing ointment. The light greenish gel fills the leaves and is very easy to extract. The juice, or gel, has been recommended for "ulcers and colon maintenance," arthritis, diabetes, and high blood pressure. In pharmacology, the dried juice has been known as "aloes" and has been used fairly extensively as a purgative, tonic, and emmenagogue.

In recent years, the gel in various forms has been incorporated in foods and beverages. For example, an Oregon firm introduced Aloe Falls, a line of six ready-to-drink teas and juice products, by way of health stores. A patent was granted to Moore and McAnalley (1996) for an aloe vera drink containing alcohol-precipitated mucilaginous polysaccharides from aloe vera juice mixed with preservatives, antioxidants, sweeteners, and flavorings — both carbonated and non-carbonated versions are describe.

Also obtained from aloe is aloin, from the outer rind of the leaf. This substance is known as a laxative, and has not yet been incorporated into a food or beverage, so far as your author knows.

Aloe is widely grown as a commercial crop, and ingredients derived from the gel are readily available, both as liquid preparations and as dehydrated powders.

Echinacea — This is a genus of plants that include purple coneflowers, has been used as a herbal remedy for infections, wounds, and snakebites by American Indians, and was fairly widely used as a cold remedy in the US during the nineteenth century. German clinical studies claim to show that Echinacea is important in strengthening the body's immune system. Herbal teas, capsules, tinctures, and other preparations including the powdered root, or extracts of the root, are available in some health stores. According to Brandt (1996), Lemon Ginger Echinacea Juice, is being marketed by a California firm. Among other ingredients are brewed ginger root tea, acerola cherry juice, and *E. purpurea* juice extract. It is described by the maker as "a healthful, juice-based beverage that appealed to cold sufferers."

Flaxseed — Flaxseed has been used at low levels as an ingredient in multigrain breads. It is one of the richest known sources of estrogen precursors and may also enhance the immune system as a result of its omega-3 fatty acid content (Best 1996).

Ginkgo — The ginkgo tree, *Ginkgo biloba*, is a popular ornamental plant. The leaves, and probably other parts of the tree, are used au naturel and as extracts. Brandt (1996) says, "Studies suggest that ginkgo improves problems associated with aging, such as memory loss and poor circulation, by improving blood and oxygen flow to the brain." It has been used in combination with other herbs and miscellaneous additives in teas, where it improves flavor not a whit. Celestial Seasonings offers GinkgoSharp Herb Tea, which is described as containing both ginkgo leaves and extract plus ginseng, golu kola, peppermint leaves, and cinnamon.

Ginseng — Ginseng can probably be found in every list of helpful herbs assembled in the last thousand years. One explanation of its values says it is regarded as "a balancer or adaptogen to make the user even-keeled; it helps the body adapt to stress and acts as an immune enhancer" (Ingeno 1996). It has also been touted as an aphrodisiac, the claim that has made the fortune of quite a few nostrum peddlers. None of these values have been substantiated by controlled scientific tests.

Ginseng root is harvested in, and enters the international market from, Siberia, Korea, China, and the US. The usual method for taking this quasi-pharmaceutical is as a tea or extract; often also in tablet or capsule form. It sometimes appears in unlikely places, as in pasta. The flavor of the product, sometimes described as mud-like, requires a learning period before its nuances are appreciated.

There are at least three species of ginseng on the market, often not distinguished on the label: *Panax ginseng* (the traditional Asian variety), *Panax quinquefolium* (grown in Canada and the US), and *Eleutherococcus senticosus* (a distantly related plant, harvested in Siberia and adjacent areas).

Ginseng-containing soft drinks have appeared in supermarkets. One of these is a carbonated beverage offered in eleven different flavors, including cola and several fruit flavors.

Green tea — This item of commerce, common green tea, has been found to contain a rich supply of phenolic catechins, potent antioxidant substances. Some of these apparently have physiological effects, such as inducing vasodilation. The enhancement of immune system reactions and cancer-fighting capabilities are also claimed.

Guarana — Originally, this was a dried paste made from the seeds of a Brazilian climbing shrub, *Paullinia cupana* that was used locally for making an astringent drink helpful in treating diarrhea. In addition to considerable tannin, this paste contains up to 4.5% caffeine. In Brazil, similar preparations have been used as an ingredient in soft drinks, including bottled carbonated beverages for popular consumption, i.e., not for treating a specific disorder.

Ingredients derived from guarana have been used in the US to provide a stimulant effect in herbal teas, tonics, and other preparations, and are believed to have been used in bottled beverages and perhaps in liqueurs.

Kava — A substance made from the roots and rhizones of two Australasian plants, *Piper methysticum* and *P. excelsum*, that has been used since prehistoric times by Maoris, and other ethnic groups, as the basis of an intoxicating beverage. Kava has been described in pharmacological texts as a diuretic and genitourinary antiseptic. Now being promoted as a sedative (said to be more potent than Valium, Halcyon, etc.), and suggested as a muscle relaxant and tranquilizer. Some attempted use in herbal teas has apparently been forestalled by legal questions, since it is clearly subject to abuse as a "recreational drug."

BIFIDIS AND ACIDOPHILUS BACTERIA

The purported health benefits of consuming yoghurt containing live bacterial cultures is responsible for selling hundreds of millions of dollars of product every year. Dairy products made with the participation of bacterial cultures are ages-old foods, e.g., cheese and buttermilk, but the concept of keeping the bacteria alive and functioning until the product is sold is relatively new. A significant negative to commercial use of such foods is that they are constantly changing during distribution and storage, so that a uniform product cannot be guaranteed to the consumer. Low temperature storage and quick turnover, conditions that can be achieved with modern practices, have made it possible to deliver to the user active yoghurts, *Lactobacilus acidophilus* milk, and similar live-culture products of dependable and relatively uniform quality.

In addition, there are dietary supplements containing freeze-dried bacteria that presumably are activated when they come into contact with moisture at moderate temperatures, and yet can be distributed without the aid of refrigeration. These products, and the aforementioned ready-to-eat foods and beverages containing live cultures have been given the name "probiotics."

There are, of course, billions of live bacteria of many different species in the lower intestinal tract of every healthy human. Not all of these are the friendly bifidobacteria (such as *Bacillus bifidus communis*) and acidophilus bacteria, however. So, "prebiotics" enter the scene. These materials "feed the beneficial microorganisms which are indigenous to the gastrointestinal tract of eaach person. From a health point of view, increasing one's own strain of bacteria is far more desirable than introducing new colonies of bacteria which may be neither compatible nor beneficial" (Anon. 1995).

There is help on the way; it is called fructooligosaccharides, a fructose analogue of the glucose polymer, starch. These materials are not digested, or are poorly hydrolyzed, by enzymes elaborated by the human alimentary apparatus. Consequently, most of the polysaccharide molecules arrive in the lower bowel substantially intact and are available to be used as a nutrient by the bifidus and acidophilus bacteria. The desirable bacteria can hydrolyze the fructose-based polymers, and use the sugars so obtained to grow and multiply.

Fructose polymers can be found in nature, as the inulin that constitutes a large part of the dry matter of the Jerusalem artichoke. These natural fruco-polysaccharides have been recommended as food ingredients or dietary supplements for increasing the population of bifidobacteria in the intestinal tract.

A more recent development is the synthesizing of fructose polymers in a controlled fashion, using sucrose as a raw material. One way of doing this is to ferment sucrose solutions with a selected strain of *Aspergillus niger* that produces a fructosyltransferase enzyme linking additional fructose molecules onto the fructose residues of sucrose.

An ingredient of this type, called NutraFlora FOS, with 95% fructo-oligosacchrides on a dry weight basis, has become commercially available within the last few years. The marketer claims (see previous citation) it is useful normalizing bowel movement, improving blood lipid composition, "inhibiting the growth of harmful E. coli and clostridia bacteria," and suppressing intestinal putrefactive substances by lowering the pH. Also, "...as little as one gram...taken regularly over a four-week period increases the amount of friendly intestinal bacteria over fivefold." Evidently, there has as yet been no particular effort made to encourage food manufacturers to use this material as an ingredient in conventional foods.

SPORTS DRINKS AND SNACKS

Athletes are partial to foods and beverages designed to give them an edge in performance or to rebuild tissues damaged by excessive activity. Less competitive citizens dedicated to workouts and jogging have come to expect the same kind of products. A large industry has been built up on the satisfaction of these real or perceived needs. The category of products is split into several different classes, but for our purposes it is helpful to consider it as being divided into sports drinks and sports snacks.

Sports Drinks

"Sports drinks" is the name given to those beverages sold primarily for the purpose of replacing both water and ions lost in perspiration during intense exercise sessions, as in jogging, football, etc. Gatorade is the tradename for the prototypical example, familiar to young and old. This is basically a flavored and colored solution of sugar — Kool Aid comes to mind. A significant difference is that many sports drinks contain a non-sweet carbohydrate, i.e., maltodextrin instead of sugar, and contain some mineral compounds designed to adjust the ionic strength of the liquid to approximate serum osmotic strength.

Why not drink water? It can be shown that, in sedentary subjects, water is absorbed more rapidly than solutions of sugar or carbohydrates. But, tests conducted with people who were exercising show that solutions of carbohydrates in concentrations up to about 8% are absorbed at the same rate as plain water (Smith 1996). At higher concentrations of sugar, absorption of water tends to slow.

Examples quoted by Smith include:

(A) When cyclists drank carbohydrate solutions before a test, they performed more work in 45 minutes than when they drank water.

(B) Cyclists given carbohydrate solutions at 12-minute intervals performed more work than when they were given water.

(C) When runners drank a 25% glucose polymer solution immediately after a run to exhaustion, their stored glycogen went up rapidly.

The sequence of physiological events detectable when a person who is exercising drinks a carbohydrate beverage is: (1) Blood sugar starts to rise after about 20 minutes; (2) Blood insulin levels increase; (3) The ratio of carbohydrate oxidation to fat oxidation increases.

One of the most famous sports drinks contains the following ingredients when in liquid form (description of function is the author's evaluation, not the manufacturer's):

Water

Sucrose — for flavor and energy

Fructose corn syrup (includes glucose) — for flavor and quick energy
Citric acid — for flavor, preservative effects, and pH adjustment
Natural flavors — probably essential oils of citrus fruits at less than 1%
Salt — to adjust ionic strength
Sodium citrate — to assist in buffering the pH at a desired level
Monopotassium phosphate — allows phosphate claim and is part of buffer
Ester gum — adjusts viscosity for increased "richness" effect
Yellow #5 — color to give the impression of citrus juice
Brominated vegetable oil — a clouding agent that improves the appearance
Yellow #6 — another coloring agent

The powder form contains sucrose, glucose, citric acid, salt, sodium citrate, natural lemon and lime flavors with other natural flavors, monopotassium phosphate, calcium silicate (an anti-caking agent) , and yellow #5.

Sports Snacks

"Sports snacks" is a fairly broad category of foods, mostly in bar form, that look suspiciously like candy or cookies until you learn they are good for you. Bars of this kind have been sold in health food stores and athletic supply centers for decades, usually under the imprimatur of a well-known athlete or trainer. Now they can be bought in any supermarket.

They have been called "ergonomic aids." Granola bars are a subdivision, but these products are now sold almost entirely on the basis of their organoleptic properties — taste, texture, and appearance — with their functional properties de-emphasized or implied by association. Chocolate-coated granola bars and granola bars with marshmallows require a leap of faith to be regarded as especially healthy foods.

The bars are usually dense and chewy and have a very non-uniform texture, full of particles of different kinds. The marketing approach is that they are energy boosters with added benefits, e.g., with added nutrients or without sucrose. Most of them combine complex carbohydrates from grains, sweeteners from fruit, vitamins, minerals, and perhaps other alleged healthy ingredients in either a taffy-like or cookie-like base. Some of them claim to increase endurance or turn fat into muscle.

According to Chase (1995), when the National Academy of Sciences evaluated six performance boosters for the US Army, they found the only ones with proven merit were carbohydrates and caffeine. Some other alleged boosters showed promise, but not at statistically valid levels. Among the additives with supposedly special values for athletes are chromium picolinate (as a fat burner and muscle builder), carnitine (as a fat burner and endurance improver), creatine (for short bursts of energy), coenzyme Q (for increased aerobic capacity), vitamin A (on the general principle`that, if a little is good, a lot is better), and vitamin E (to prevent oxidative damage in weightlifters and downhill runners).

Formulas and processes — Formulas for sports snacks must be based on the expectations and beliefs of the specialized customer group. The basic structure is generally formed by gently mixing the particulate material, such as nuts, whole or crushed grains, and raisins or other fruit pieces, with hot syrup containing the liquid or soluble ingredients. Obtaining a mixture uniform enough to insure that the ingredient statement applies to every sale portion (bar, etc.) is difficult in such cases, partly because of the predominance of relatively large particles of several different kinds. It follows that manufacturers prefer smaller particles to larger particles and that a very uniform pre-mix of the solid material should be made before the syrup is added — precautions should be taken against grading (separation by particle size or density) when mixing, transferring, and weighing.

Forming of individual bars can be done manually by spreading the warm and still plastic mixture into rectangular pans or trays, rolling or pressing to achieve a uniform thickness, then cutting (or pressing with a sharp die) to establish the length and width dimensions. Mass production can be achieved by spreading the warm mixture onto a conveyor belt, passing the mixture under a series of rollers to establish a uniform and predetermined thickness, then cutting pieces with rollers having die blades arranged to make the width and length cuts separately — alternatively, the plastic mixture can be pressed into dies. The adhesive nature of most of these mixtures will present problems of die release and conveyor fouling at every step. In some cases, the finished bar will will require drying to remove a few percent of water; generally, mild methods suffice to achieve this level of dehydration. In other cases, absorption of moisture by the dry ingredients is sufficient to achieve the desired texture while maintaining the low water activity needed for preservation.

Bars can also be made by preparing a wetter but cohesive premix (a dough) and baking to form a solid structure, analogous to making bar cookies by the well-known continuous methods for which equipment is readily available. Batch methods are also suitable for this type of product. Baking conditions should be fairly mild and should not cause significant deterioration of the nutrients, but this presumption must be verified for each individual formula.

NONCARIOGENIC FOODS AND BEVERAGES

The general impression given by current publications, advertising, and label statements is that choosing foods on the basis of their value in preventing tooth decay is no longer a major priority of the American consumer. It is true that a few food labels can be found that do include claims of caries prevention or non-cariogenicity.

Generally, the approach in achieving these properties is to reduce or eliminate the amount of sugar in a portion of food. There is evidence that sucrose becomes a nutrient for specific bacteria inhabiting the mouth and that the insoluble slucan produced by the bacteria facilitates the formation of plaque, which seems to be a prelude to destruction of enamel. Removing sucrose, or other carbohydrates usable by bacterial flora always present in the buccal cavity, from conventional formulas and replacing them with non-utilizable carbohydrates plus, perhaps, high potency sweeteners, should reduce tooth decay. One of many patents following this line of reasoning will serve as an example. Iijima et al. (1990) disclosed a sweetened condensed milk replacement made by using a palatinose syrup and crystallized palatinose as a sweetener instead of the conventional sucrose, palatinose being a sugar that does not contribute to the formation of glucan and plaque.

Foods having a sticky character, or that leave a residue on the teeth because of some other property, are particularly damaging. In a classic study in Sweden, caramel candy that contained only 17% to 18% sucrose produced a higher level of tooth cavities than other less adhesive foods containing much more sucrose. Dried fruits, such as raisins, are also particularly harmful. There might be a market for modified products of these types that, somehow, eliminate or reduce their cariogenic properties. Caramel candy made with the palatinose syrup described above, or in which the content of fermentable sugars has been greatly reduced by some other method, might be one approach. Alternatively, taffy or caramel that is not sticky could perhaps be devised; it is unlikely that stickiness itself is one of the desired organoleptic properties of these products, being instead an unavoidable result of formulation and processing details introduced to achieve other ends.

Sometimes, advantages are said to be gained from tough, chewy, or abrasive foods that apparently scour bacteria from the surface of teeth or remove plaque. In a few cases, foods are said to contain specific compounds that by some poorly understood mechanism combat tooth decay directly. Madsen (1981) pointed out that finely ground dietary oat hulls inhbit caries under many different assay conditions. Both smooth-surface caries and fissure caries are decreased.

Anticaries activity has been found in the seeds of ten other species: the hulls of rice, barley, cottonseed, cocoa, and soybeans, the shells of peanuts and pecans, the bran and polishings of rice, sunflower hulls, and corn are active in inhibiting caries. A part of this effect may be due to chemicals present in the hulls, bran, etc., that are not found in appreciable amounts in the other portions of the seed. For example, it was noted by Madsen that phenolic acids and polyphenols are present in the hulls.

HUNGER SUPPRESSANT FOODS AND BEVERAGES

In a patent granted in 1996 and assigned to Pepsico, Kwapong and Fedunjacklin (1996) described a method of producing satiety by increasing the titratable acidity of a foodstuff, particularly beverages. A conventional tasting soft drink can be made even when the titratable acidity is raised to 125. The novelty consists in using a blend of common acids, more or less in accord with the practice in most carbonated soft drinks, but in greater quantity, and then adding glucono-delta-lactone which helps to smooth the acidic flavor and produces a more acceptable beverage. The key appears to be that a high level of total acid dulls the appetite while the glucono-delta-lactone makes the extreme tartness or sourness of the beverage less noticeable.

There is some evidence that 2,5-anhydro-D-mannitol is an appetite suppressant, at least in some animals. The previous history of this concept is review in the patent of Friedman et al. (1989), which discloses a method for altering the food intake of mammals by administering small amounts of this fructose analogue at specific intervals. Although the regulatory status of the compound is not known by the author, it would appear to have some potential for medical treatment of refractory obesity.

Also of interest is the high fiber powder described as a weight-control formulation (Gori 1988). The inventor describes this material as "A packaged weight-control powder . . . to be employed before eating. Each pouch includes a dose of about one gram to be sprinkled on the foodstuff . . . The powder is formed of a mixture of oat, wheat, and corn brans mixed with pectin guar gum, psyllium, and cutin. Mineral supplements are added to replace those removed by the fibers of the brans." The mixture is sprinkled on prepared food, e.g., mashed potatoes. About ten packets to be used each day. The material functions as a weight control aid, according to the inventor, because it absorbs water and gives a sensation of satiety with less calories.

30 grams oat bran
15 grams wheat bran
10 grams corn bran
10 grams pectin, technical grade
20 grams guar gum
10 grams psyllium seed
 5 grams cutin (apple skins)
10.0 mgs manganese
 0.1 gram selenium
20.0 mg zinc
 5.0 mg copper
20.0 mg iron
200.0 mg calcium
150.0 mg magnesium

SUPPLEMENTS

There have been continuing publicity, political, and legal battles between those persons who wish to market concentrated and/or specialized nutrients (as defined by them) without legal interference of any kind and governmental agencies (and some medical associations) who wish to rigidly enforce existing laws, by (for example) classifying the substances as drugs/medicines/pharmaceuticals, a move which would, in fact, result in the outlawing of most of this traffic. As is usual in politico-economic issues of this type, each side has had a certain amount of success in getting their point of view across, and an intermediate position of regulation is in effect at the present time.

One type of product that is allowed, if it conforms to certain standards, is the dietary supplement tablet, capsule, powder, infusion, or mixture containing relatively large amounts of vitamins, minerals, and proteins (or amino acids), sometimes with flavors or materials of questionable dietary significance. The type of materials we will be considering in this section are known collectively as "supplements."

What are Supplements?

The group of ingestibles now called "Supplements" are substances that fill the gap between, and overlap, the categories of foods and orally applied medications. There is an enormous range of products being offered in this group; they are are sold in the form of capsules, tablets, liquids, powders, gums, gels — almost every imaginable physical form. They include proteins, amino acids, vitamins, minerals, extracts of animal and vegetable materials, fish oils, soluble fibers such as gums and mucilages, and substances synthetically produced or of natural origin not generally recognized as nutrients, such as bioflavonoids, enzymes, germanium, nucleic acids, and rutin.

Typical examples of supplements are vitamin and mineral tablets, garlic pills, ginseng tonics, and herbal teas. The manufacturers and sellers of supplements have tried, with considerable success, to keep the products from being regulated as either foods or medicines since, under the current methods of production, promotion, sale, and use, it is very clear that the great majority of the items could not meet the current regulatory requirements for either group.

The reason is: for the majority of the supplements, there has never been any scientific study showing any nutritive or medical benefit for humans from the consumption of the items. Indeed, many supplements are undoubtedly potentially harmful to at least some people. Nonetheless, the explicit or implied reason for buying virtually all the ingesta found in, for example, a General Nutrition store, is health improvement, as opposed to the organoleptic attractions of "ordinary" foods and beverages.

How Supplements are Regulated

Under the Federal Food, Drug, and Cosmetic (FD&C) Act, FDA regulates dietary supplements as foods, so long as no drug claims are made for them. The Act includes provisions intended to ensure that foods are safe and their labels are truthful and not misleading. The Proxmire Amendment of 1976 restricts FDA's authority to limit supplement potency and the composition of most multi-nutrient products, except for safety reasons. It prohibits the agency from treating a vitamin or mineral as a drug solely because of potency.

When products are marketed for therapeutic use, FDA regulates them through its Center for Drug Evaluation and Research. The FD&C Act defines "drugs" as products intended to diagnose, cure, mitigate, treat, or prevent disease, or non-food products intended to affect the structure or a function of the body.

The manufacturers, distributors, and retailers of supplements have a powerful political influence. They managed to get enacted into law the Dietary Supplements Health and Education Act (DSHEA) of 1994, which was intended to relax the regulations affecting their group of products. It provided a separate regulatory category distinct from foods and medications, the differences being particularly evident in the labeling details. In general, the act has had the expected effect of expanding the opportunities for new products and reducing the need for proof of effectiveness and safety of the category, but it may have also made it more difficult to market the phytochemical benefits of food products (Allen 1996B).

Final standards are still pending as of this writing, but the current version defines a *dietary supplement* as follows: "A product that is intended to supplement the diet that bears or contains one or more of the following dietary ingredients: a vitamin, a mineral, an herb or other botanical, an amino acid, a dietary substance for use by man to supplement the diet by increasing the total daily intake, or a concentrate, metabolite, constituent, extract, or combinations of these ingredients." And:

• It is intended for ingestion in pill, capsule, tablet, or liquid form.

• It is not represented for use as a conventional food or as the sole item of a meal or diet.

• It is labeled as a "dietary supplement."

• It includes products such as an approved new drug, certified antibiotic, or licensed biologic that was marketed as a dietary supplement or food before approval.

The 1994 law permitted the use of a "structure/function claim," allowing a reference to the nutrient's function in the body or its effect on the structure of the body. Unlike a health claim, it does not link a product or substance to the cure or prevention of a specific disease or health condition. Notification of the FDA is required **after** the product enters the market, a very significant difference from the case with new food products or ingredients intending to be associated with health claims.

According to one marketer of ingredients and products supposedly protected and legitimated by the DSHEA, "Herbal products were never covered as dietary supplements under the Code of Federal Regulations . . . In the past, if you put label claims on your package, the product was considered a drug, but under the DSHEA, companies can make structure and function claims as long as the label clearly indicates the product is a dietary supplement and not a food." This also means that manufacturers must substantiate label claims. If the claims are incorrect, due to missing ingredients or ingredients present at less than the declared level, the FDA can prosecute. Brandt (1996) gives an example: If ginkgo tea is sold as a supplement, the manufacturer can say, "Gingko extract supports normal oxygen flow to the brain," which Brandt says has been scientifically proven — however, manufacturers cannot state that the product itself will cure, prevent, or treat a disease.

The FDA is currently proposing that the words "Dietary Supplement" be required on the front panel of a product package and that nutrition information be listed under "Supplement Facts."

Some differences between the labeling of supplements and foods are:

1. Supplement labels have a different title than food labels.

2. Serving size is designated; e.g., one packet, two tablets, two cookies.

3. All nutrients are listed on a per-serving basis.

4. Weight listed is weight of *ingredient* (i.e., "nutrient") not weight of *source* (raw material or as-is ingredient).

5. Only nutrients present in significant amounts can be listed.

6. Ingredients only need to be listed once; if not listed in the principal table, they should be listed below the table as "Other ingredients: . . . "

7. Supplement label allows listing both quantity and percent DV.

8. Quantities are rounded just like food products.

9. Daily values (DVs) are rounded to nearest percent.

10. Botanicals are listed by common name and Latin botanical name of the plant, and part of plant used (leaf, root, flower, etc.).

11. Must use correct units of measure, as shown in DV chart (Table 13.1).

12. The source of the nutrient may be listed in parentheses next to or immediately under the name; "as" or "from" is also required.

13. Nutrients that do not have an assigned Daily Value may be put on the label, but are set apart.

Table 13.1
DAILY VALUES OF VITAMINS AND MINERALS[1]

Vitamin A	5000 IU
Vitamin C	60 mg
Calcium	1000 mg
Iron	18 mg
Vitamin D	400 IU
Vitamin E	30 IU
Vitamin K	80 mcg
Thiamin	1.5 mg
Riboflavin	1.7 mg
Niacin	20 mg
Vitamin B_6	2.0 mg
Folate	400 mcg
Vitamin B_{12}	6 mcg
Biotin	300 mcg
Pantothenic acid	10 mg
Phosphorus	1000 mg
Iodine	150 mcg
Magnesium	400 mg
Zinc	15 mg
Selenium	70 mcg
Copper	2.0 mg
Manganese	2.0 mg
Chromium	120 mcg
Molybdenum	75 mcg
Chloride	3400 mg

[1] Daily values are in the order they should appear on the label.

GLOSSARY

A collection of terms used in this book that may not be familiar to food technologists and which may not be defined in the text of this book.

adult-onset-diabetes — a relatively mild form of diabetes mellitus that usually develops in persons over 35.

arteriosclerosis — abnormal thickening and hardening of the arteries, especially of the intima, occurring mostly in old age.

atheroma — a soft encysted tumor containing curdy matter. A disease characterized by fatty degeneration of the inner coat of the arteries.

atheromatous — of, pertaining to, or having the nature of, atheroma.

atherosclerosis — a type of arteriosclerosis in which fibrous thickening of the intima is accompanied by atheromatous degeneration.

beta-glucans — long chain glucose polymers differing from cellulose in that they contain beta 1-3 links as well as beta 1-4 links, in various proportions depending on the source.

bifidogenic — promotes growth of (specifically) bifidis bacteria in the lower intestine.

bioactive — having some supposedly beneficial effect on the functioning of the body.

body mass index also, BMI. A measure frequently used in reports of body weight problems; consists of the ratio of weight (in kilograms) to the square of height (in meters).

cellulose — a linear polymer of glucose connected by beta 1-4 links; the main structural component of plant cell walls.

CFR — Code of Federal Regulations. A many-volumed compendium of essentially all the regulations issued by the government. Citations are given in the form 21 CFR 86.1, where the first 1 or 2 numbers refer to the volume, and the last figures to the section.

cis — [usually in italics] a prefix indicating that certain atoms or groups are on the same side of the molecule.

colligative — *Physical Chem.* Depending on or varying with, the number of molecules and not their nature; as gaseous pressure is a colligative property.

colonic food — (n) a food ingredient/component that, due to resistance to digestion, reaches the colon substantially intact, where it is fermented by, and influences, the composition/activity of the colonic microbiota and consequently the physiology of the colon.

complex carbohydrates — not currently a description accepted by the FDA for labeling purposes, though they once defined it as the sum of dextrins and starches. There is a proposal to define the term as meaning the sum of starch and dietary fiber. In the UK, generally defined as the sum of starch and non-starch polysaccharidesin a food.

coronary arteries — the right and left arteries that arise from the aorta immediately above the semilunar valves and supply the tissues of the heart itself with oxygen and nutrients.

coronary heart disease — disorders of the coronary arteries.

cutin — a complex mixture of waxes, fatty acids, soaps, higher alcohols, and resins that forms the chief component of the cuticle of many plants.

cyclohexyl diol esters — fatty acid diesters of cyclohexanediol, cyclohexenediol, and cyclohexdienediol, and their dimethanol and diethanol counterparts

demography — the statistical study of populations, as to births, marriages, mortality, health, etc.; — usually restricted to physical conditions or vital statistics.

designer foods — a research concept dealing with the food-disease prevention value of non-nutritive secondary plant metabolites in the human diet.

dietary fiber — (n) the sum of non-starch polysaccharides and lignin that is not digested by human gastrointestinal enzymes, but may be partially degraded by colonic bacteria. Other definitions have been used, and may be in use.

diverticular disease — an abnormality that can occur anywhere along the colon, but most frequently in the sigmoid colon. Abnormal pressure over many years weakens areas in the muscular wall of the colon. Herniations or formation of extruding pouches may develop.

DSHEA — the Dietary Supplement Health and Education Act of 1994.

enrichment — the addition of nutrients to achieve in the finished food concentrations specified by Standards of Identity, or other specifications. The word is often used as a synonym for "fortification."

enteral formulas — nutritional preparations used to feed hospitalized patients who require special physical forms of or nutrient combinations in their ingesta, e.g., for liquid diets; products fed by nasogastric or intestinal tubes also fall into this category. Opposed to parenteral fluids, which are delivered into veins of patients. cecum for defecation.

fat — substances found in animal and vegetable tissues that are, in a chemical sense, glyceryl esters of fatty acids; generally understood to be triglycerides incorporating naturally occurring fatty acids. Fats yield an average of about nine calories per gram in bomb calorimetry measurements, but in physiological terms, they vary in calories depending upon the structure of the fatty acids.

folate — an orange crystalline vitamin $C_{19}H_{19}N_7O_6$, of the B group, The chemical name is pteroylglutamic acid, and it has also been called Vitamin B_c, vitamin L, vitamin M, and lactobacillus casei factor. In nature, it is found in relatively high concentration in some leaves; it is now prepared synthetically. It has been used in the treatment of macrocytic anemia. The form in which folic acid is found in nature.

food — 1. Nutritive material absorbed or taken into the body of an organism which serves for purposes of growth, work, or repair and for the maintenance of the bodily processes. 2. Nutriment in solid form, as opposed to "drink," or "beverages," which may also contain nourishing substances.

fortification — the addition of nutrients at levels higher than those found in the original (or a comparable) food.

free radical — a compound or element with an unpaired electron, generally unstable and very reactive. Carbonyl, CO, nitroso radical, NO, the sulfonyl group, SO_2, and even some complex organic compounds, e.g., triphenylmethyl, are free radicals.

functional food — a food that affects physiological functions of the body in a targeted way so as to have positive effects on health.

gastric emptying — movement of stomach contents into the duodenum.

glucose tolerance test — oral intake of a precise amount of glucose to measure plasma glucose changes at periodic intervals as a means of determining the ability to utilize the sugar.

glucosuria — an abnormal condition in which glucose appears in the urine.

GRAS — generally recognized as safe; a legal way of sidestepping the lengthy and expensive FDA approval process required for new foods and ingredients; in essence, a grandfathering of a material by claiming that it has been in use for many years without reported harmful effects. There are thousands of GRAS ingredients.

gums — any of a class of colloidal substances, glutinous when moist but hardening on drying, exuded by plants or extracted from them by solvents, and soluble in (or swelling up with) water, consisting mostly of complex organic acids or their salts. Also, synthetic materials having similar properties.

HDL-cholesterol — cholesterol in the blood that is transported by high-density lipoproteins (HDL); high concentrations are thought to be beneficial.

health claim — (FDA) any claim made on labels or in labeling of food, including a dietary supplement, that expressly or by implication, including "third party" endorsements, written statements (for example, a brand name including a tgerm such as "heart"), symbols (for example, a heart symbol) or vignettes, characterizes the relationship of any substance to a disease or health-related condition. Implied health claims include only those statements, symbols, vignettes, or other forms of communication that a producer intends, or would be likely to be understood, to assert a direct beneficial relationship between the presence or level of anything in the food and a health- or disease-related condition.

healthy — according to FDA, conditions for the use of "healthy," a food must meet the definition of "low" for fat and saturated fat, and neither cholesterol nor sodium may be present at a level exceeding the disclosure levels in 21 CFR 101.13(h). In addition, the food must comply with definitions and declaration requirements for any specific nutrient content claims.

hemicellulose — any of a group of plant polysaccharides that are extractable by aqueous alkali; they resemble cellulose in some respects, such as being insoluble in water, but they are less complex and more easily hydrolysable to simple sugars. Those yielding pentoses on hydrolysis are called pentosans, those yielding hexoses are called hexosans.

homocysteine — an amino acid that is a homologue of cysteine, and is produced when methionine is demethylated; it forms a complex with serine that is metabolized to yield cysteine and homoserine..

hypercholesterolemia — a condition in which the concentration of cholesterol in the blood is abnormally elevated.

hyperglycemia — a condition in which the concentration of glucose in the blood is abnormally high, especially with reference to the fasting level

hyperlipidemia — a general term describing the condition in which the concentration of lipids (cholesterol and/or triglycerides) in the blood is abnormally elevated.

IgE — immunoglobulin E, one of the antibody systems that the human body uses to fight infection and resist disease, and which mediates the untoward results of food allergenicities.

intima — the innermost coat of an organ, especially of a blood vessel or lymphatic; in the larger blood vessels it consists of an endothelial lining backed by a layer of connective tissue and one of elastic tissue.

inulin — a nearly tasteless, white, semi-crystalline polysaccharide, composed primarily of fructose residues, that can be isolated from the sap of the roots and rhizomes of many plants. It resembles starch in many of its properties. Essentially non-digestible.

LDL-cholesterol — the part (usually about three-fourths) of the blood's cholesterol that is transported by low-density lipoproteins (LDL); it is the major fraction that becomes elevated in hypercholesterolemia.

lignin — a complex mixture of non-polysaccharide polymers of plant alcohols that is related physiologically to cellulose and with it constitutes the essential part of woody tissue; it contains phenylpropane units derived from phenolics such as sinapyl, coniferyl, cinnamyl, and p-coumaryl alcohols. Lignins are insoluble and resist digestion in the alimentary tract.

lipid — also, lipide. Any of a group of substances comprising the fats and other esters that possess analogous properties. They are characterized by solubility in chloroform and other solvents for fats, insolubility in water, and by their greasy feel. They have been divided into simple, or ternary, lipids, which contain only carbon, hydrogen, and oxygen (glycerides, cerides, sterides, and etholides), and complex lipids which also contain phosphorus (phospholipids) or phosphorus plus nitrogen (phosphoaminolipids).

lipoprotein — any of a cla   of conjugated proteins containing a lipid group. The chemical form in which cholesterol is transported in the blood.

meal tolerance test — measuring the plasma glucose changes at set intervals after intake of a precise amount of carbohydrates along with other nutrients in a defined meal.

medical foods — enteral formulas (q.v.).

mineral bioavailability — the percent of the intake of a "mineral" (e.g., iron, calcium, phosphorus) that is absorbed and retained by body tissue for biological function.

mucilages — gelatinous substances found in plants, as in fucoid seaweeds, flaxseed, marshmallow, and quinces. Chemically, they are mixtures from which have been isolated complex carbohydrates (such as araban), and related compounds (such as alginic acid).

NLEA — the Nutritional Labeling and Education Act of 1990.

nutraceutical — also "nutriceutical." A substance considered a food, or part of a food, with medical or health benefits, including the prevention, treatment, or cure of disease.

oat bran — the food produced by grinding clean oat groats or rolled oats and separating the resulting oat flour by sieving, bolting, and/or suitable means into fractions such that the oat bran fraction is not more than 50% of the starting material, and has a total beta-glucan content of at least 5.5% (dry weight basis) and a total dietary fiber content of at least 16.0% (dry weight basis), and that at least one-third of the total dietary fiber is soluble fiber.

omega-3 fats — fats containing polyunsaturated fatty acids that have the first double bond positioned after the sixth carbon atom from the methyl (CH_3 end of the chain). The fatty acids in which this occurs are eicosapentenoic and docosahexaenoic. In the foods category, fish oils are found to contain significant amounts of these fatty acids.

orthomolecular — related to or aimed at reaching optimal concentration and functions at the molecular level of the substances normally (or correctly) present in the body.

pectic substance — any of a group of complex colloidal carbohydrate derivatives of plant origin, containing a large proportion of units derived from galacturonic acid. The chief classes are protopectins, pectins, and pectinic and pectic acids.

pectin — any of the group of water-soluble, methylated pectic substances occurring in plant tissues or obtained by restricted treatment of protopectin with protopectinase, acids, or other reagents; also, the group as a whole.

performance food — (n) essentially the same as "functional food," but sometimes considered to be restricted to those foods or supplements that promote some specific physical (even at times, mental) trait, such as endurance or strength.

pharmafoods — nutraceuticals.

phytochemical — any chemical produced by a plant, but often used to mean plant constituents that are bioactive.

plasma cholesterol — the total concentration of cholesterol expressed as mg per 100 ml of plasma (the liquid portion of uncoagulated blood, containing fibrinogen, remaining after removal of the red blood cells).

polysaccharide — any carbohydrate decomposable by hydrolysis into two or more molecules of simple sugars or monosaccharides; specifically, one of those of more complex composition $(C_6H_{10}O_5)_x$, formerly called amyloses, as starch and cellulose.

postprandial — following a meal,

prebiotic — a non-digestible ingredient beneficially affecting the host by selectively stimulating the growth or activity of one or a limited number of bacterial species in the colon.

probiotic — foods or supplements containing live bacteria of the type thought to be beneficial in the lower intestine (such as bifidis); typically, acidophilus milk.

protein efficiency ratio — also, PER. An indicator of the proportions of essential amino acids present in a food as compared to the body's requirements.

proximate analysis — 1. Determination of compounds, or classes of compounds, moisture, ash, etc., in a complex substance. 2. Determination of radicals in a compound.

regimen — a regulation or treatment intended to benefit by gradual operation, such as a systematic course of diet to improve or preserve health, or to attain some effect, as the reduction of body weight.

resistant starch — a substance having the chemical structure of normal starch (amylose or amylopectin) but, due to a difference in its physical structure, is not digested, or is incompletely digested, in the human alimentary tract.

restoration — when applied to nutrient additions to foods and food ingredients, this term means the replacement of certain nutrients lost during processing, so that the finished food contains as much of the nutrient as was present in the raw material(s).

reversion — off odors and flavors in refined oils that cannot be accounted for by any measure of oxidation. Certain oils, and particularly soybean oil, is more subject than others to the development of grassy, painty, and other undesirable odors upon aging. Reversion can be prevented by proper processing.

saponins — any of a group of glucosides occurring in many plants, characterized by their property of producing a soapy lather; generally bitter and acrid in taste, and some are poisonous. On hydrolysis, they yield sugars and sapogenins.

short-chain fatty acids — fatty acids having six or fewer carbon atoms in their structure. Acetic acid, propionic acid, and butyric acid are examples.

sitosterol — a colorless crystalline sterol found in wheat embryos, Calabar beans, and elsewhere. There are several varieties.

soy fiber — in trade terms, the dietary fiber derived from dehulled and defatted inner parts of soybean seed (cotyledon); not soybean hull or soy bran. Composed mostly of polysaccharides.

sterol — any of a class of higher alcohols, as cholesterol and sitosterol, widely distributed in nature. They are, in general, colorless crystalline compounds, nonsaponifiable, and soluble in certain organic solvents.

tofu — a cheese analogue made from coagulated soybean protein, generally white or off-white, soft, and slightly elastic; seldom if ever aged for flavor development.

trans — [usually in italics] a prefix indicating that certain atoms or groups are on opposite sides of the molecule.

trans fat — triglyceride in which one or more of the fatty acids are of the *trans* configuration.

transit time — the time required for food to be transported from mouth to cecum for defecation.

BIBLIOGRAPHY

Some titles have been truncated, abbreviated, and/or summarized to save space. Conventional abbreviations have been used for periodical titles. If more than three authors appeared on the publication, we have used only the first with "et al."

Aaron, J., and Stauffer, C. 1986. Dietary fiber: Accessing its ingredient function and marketing role. Baking Industry *153*, No. 4, 46.

Adams, M. 1983. Salt. Proc. Am. Soc. Bakery Engrs. *1983*, 152.

Al-Qaradawi, Y. 1984. The Lawful and the Prohibited in Islam. The Holy Quran Publishing House, Beirut, Lebanon.

Allen, A. 1995A. Cashing in on the kosher cachet. Food Product Design *5*, No. 6, 69.

Allen, A. 1995B. Getting organic: What to do before the National Standard. Food Product Design *5*, No. 2, 105.

Allen, A. 1995C. Translating mixed signals on trans fat. Food Product Design *5*, No. 8, 30.

Allen, A. 1996A. Big news about beta-carotene: Stick to dietary sources. Food Product Design *5*, No. 12, 14

Allen, A. 1996B. DSHEA:One year after. Food Product Design *5*, No. 11, 16.

Allen, A. 1996C. Olestra approval focuses spotlight on regulatory dilemma. Food Product Design *5*, No. 10, 17.

Allen, A. 1996D. Phytochemical future: Can health benefits outweigh the regulatory challenges. Food Product Design *5*, No. 12, 88.

Ang, J., and Miller, W. 1991. Multiple functions of a powdered cellulose as a food ingredient. Cereal Foods World *36*, 558.

Anon. 1982. Everything kosher? Not quite . . . Snack Food *17*, No. 2, 32.

Anon. 1985. Report of Expert Advisory Committee on Dietary Fibre. Health Protection Branch, Health and Welfare Dept., Canada Information Letter *700*.

Anon. 1987A. Product data: Low moisture apple and pear fibers. TreeTop, Selah, WA

Anon. 1987B. Tolerating lactose. Calcium Currents *3*, No. 2, 1.

Anon. 1988A. Manganese lack may lead to osteoporosis. Chem. Eng. Nes *64*, No. 37, 5.

Anon. 1988B. The atherogenic potential of foods. Nutrition Reviews *46*, No. 9, 313-315.

Anon. 1988C. The role of sodium vs. sodium chloride in hypertension. Nutrition Reviews *46*, No. 5, 187-188.

Anon. 1989A. Diet and Health Implications for Reducing Chronic Disease Risk. National Academy Press, Washington, DC

Anon. 1989B. Dietary fiber. Food Technology *9*, No. 10, 133.

Anon. 1990. Rice bran's growing bakery applications. Rice Bran News *2*, No. 4, 1.

Anon. 1992A. Medical foods. Food Technology *46*, No. 4, 87.

Anon. 1992B. Solka-Floc reduces calories and improves pasta extrusion. Solka-Floc at Work *1*, No. 2, 1.

Anon. 1992C. The Solka-Floc system reduces calories in sweet baked goods. Solka-Floc at Work *1*, No. 1, 1.

Anon. 1993. The Nutrition and Labeling Education Act and the Benefits of Using Solka-Floc. Solka-floc at Work *2*, No. 1, 1.

Anon. 1994A. Food allergy databases gain in Europe. Prepared Foods *163*, No. 7, 49.

Anon. 1994B. Guilt-free cheesecake. Prepared Foods *163*, No. 9, 131.

Anon. 1994C. Trimming the fat. Baking & Snack *16*, No. 4, 23.

Anon. 1994D. Guilt-free cheesecake. Prepared Foods *163*, No. 9, 131.

Anon. 1994E. Two thousand times as sweet as sugar. The World of Food Ingredients *1994*, October/November, 59.

Anon. 1995. Inside the cultural revolution: Aiding your immune system with NutraFlora FOS Prebiotic. Nutrition Science News (July 1995), New Hope Communications, Westminster, Colorado.

Anon. 1996A. Fat Reduction in Foods (Revised). Calorie Control Council, Atlanta GA.

Anon. 1996B. Thaumatin — the sweetest substance known to man has a wide range of food applications. Food Technology *50*, No. 1, 74.

Athanasios, A., and Templeman, G. 1992. Process for the removal of sterol compounds and saturated fatty acids. US Pat. 5,091,117.

Awad, A., and Smith, D. 1996. Cholesterol-reduced egg products. U.S. Pat. 5,484,624.

Babayan, V. 1989. Sense and nonsense about fats . . . Food Technology *43*, No. 1, 90.

Babayan, V. 1991. Medium-chain triglycerides. Cereal Foods World *36*, 793.

Baer, C., et al. 1991. Low calorie food products having smooth, creamy, organoleptic characteristics. U.S. Pat. 5,011,701.

Bailey, L. 1988. Factors affecting folate bioavailability. Food Technology *42*, No. 10, 206.

Bakal, A. 1994. The lowdown on formulating lowfat. Prepared Foods *163*, No. 8, 75

Barndt, R., and Jackson, G. 1990. Stability of sucralose. Food Technology *44*, No. 1, 62.

Baughman, M. 1988. Killer cholesterol? Industrial Chemist *9*, No. 4, 7.

Baum, R. 1996. Olestra and the importance of FDA. C&EN *74*, No. 20, 25.

Beauchamp, G. 1987. The human preference for excess salt. Am. Scientist *78*, No. 1, 27.

Bedenk, W., and Purves, E. 1972. Production of high protein ready-to-eat breakfast cereals containing soy and malt. U.S. Pat. 3,682,647.

Beitz, D. 1995. Cholesterol reduction. US Pat. 5,436,004.

Best, D. 1989A. A fiber is not a fiber is not. Prepared Foods *158*, No. 3, 91.

Best, D. 1989B. High-intensity sweeteners lead low-calorie. Prepd. Foods *158*, No. 3, 97.

Best, D. 1989C. Processors pursue perfect nutritional profile. Prepd. Foods *158*, No. 3, 79

Best, D. 1989D. Unsaturated fats: know your source well. Prepared Foods *158*, No. 3, 87.

Best, D. 1993. Food technology faces the future. Prepared Foods *162*, No. 6, 32.

Best, D. 1996A. A functional pharmacopoeia. Prepared Foods *165*, No. 4, 38.

Best, D. 1996B. Nutraceuticals suit up to play. Prepared Foods *165*, No. 1, 33.

Bettschart, A. 1991. Dietary fiber in cereal grains. Rice Bran News *3*, No. 1, 3.

Beveridge, J., Haust, H., and Connell, W. 1964. Magnitude of the hypocholesterolemic effect of dietary sitosterol in man. J. Nutrition *83*, 119.

Beyts, P. 1990. Sweetener for beverages. US Patent 4,915,969. disease. Wall Street J. *81*, No. 120, 1.

Blackburn, G. 1994. Treatment & prevention of human obesity. Food Tech. *48*, No. 2, 13.

Blankers, I. 1995. Properties and applications of lactitol. Food Technology *49*, No. 1, 66.

Bogdanich, W. 1987. Panel is seeking tests for cholesterol. Wall Street J. 8 June 1987, 22.

Bosley, G., and Hardinge, M. 1993. Seventh-day Adventists: dietary standards and concerns. Food Technol. *46*, No. 10, 112.

Bostom, A. 1996. Folic acid fortification of food. J. Am. Med. Assoc. *275*, 681.

Brandt, L. 1996. Using exotic ingredients. Food Product Design *6*, No. 1, 54.

Bringe, N., and Cheng, J. 1995. Low-fat, low-cholesterol egg yolk in food applications. Food Technology *49*, No. 5, 94.

Brooks, E. 1994. Risk-free foods: Half-baked thinking. Food Technology *48*, No. 11, 22.

Browne, M. 1993. Label Facts for Healthful Eating. Nat. Fd. Proc. Assn., Washington, DC

Bruinsma, B. 1991. Fiber update — Part I. Am. Soc. Bakery Engrs. Bull. *223*.

Bucher, H., et al. [six other co-authors] 1996. Effects of dietary calcium supplementation on blood pressure. J. Am. Med. Assoc. *275*, 1016.

Bullock, L., et al. 1992. Replace simple sugars in cookie doughs. Food Tech. *46*, No. 1, 82.

Burkitt, D. 1973. Some disease characteristics of modern Western civilization. Brit. Medical J. *1*, 274.

Camire, M. 1996. Blurring the distinction between dietary supplements and food. Food Technology *50*, No. 6, 160.

Campbell, L., Ketelsen, S., and Antenucci, R. 1994. Formulating oatmeal cookies with calorie-sparing ingedients. Food Technology *48*, No. 5, 98.

Caragay, A. 1992. Cancer-preventive foods, ingredients. Food Technology *46*, No. 4, 65.

Cardozo, M., and Eitenmiller, R. 1988. Total dietary fiber analysis of selected baked and cereal products. Cereal Foods World *33*, No. 5, 414.

Carr, J. 1993. Hydrocolloids and stabililizers. Food Technology *47*, No. 10, 100.

Carr, J., Sufferling, K., and Poppe, J. 1995. Hydrocolloids and their use in the confectionery industry. Food Technology *49*, No. 7, 41.

Chalup, W., and Sanderson, G. 1996. Preparing low-fat fried-type or baked food products. US Pat. 5,492,707

Chase, M. 1995. Sports snacks gain popularity claims still studied. WSJ *96*, No. 35, B1.

Chaudry, M. 1992. Islamic food laws: philosophical basis and practical implications. Food Technol. *46*, No. 10, 92.

Chaudry, M., and Regenstein, J. 1994. Trends in Food Sci. & Technology *5*, 165.

Cleave, J., Campbell, G., and Painter, N. 1969. Diabetes, Coronary Thrombosis and the Saccharine Diseases. John Wright & Sons, London.

Colliopoulos, J. 1991. Psyllium mucilloid fiber food products.U.S. Pat. 5,009,916.

Conte, J., Jr., Johnson, B., Hsieh, R., and Ko, S. 1992. Method for removing cholesterol from eggs. US Pat. 5,091,203.

Conway, R. 1994. Good science drives quest for quality in field of nutritional analysis. Food Science Newsletter (Hazleton Labs) No. 63.

Cook, J., Minnich, V., Moore, C., Rasmussen, A., Bradley, W., and Finch, C. 1973. Absorption of fortification iron in bread. Am. J. of Clinical Nutrition *26*, 861.

Cooper, H. 1996. Scientists study how some centenarians have managed to stay hale and hearty. Wall Street J. *97*, No. 96, B1.

Cowley, A., Jr., and Roman, R. 1996. The role of the kidney in hypertension. J. Am. Med. Assoc. *275*, 1581.

Cunnane, S., et al. 1993 High alpha-linolenic acid flaxseed: Some nutritional properties in humans. Br. J. Nutr. *69*, 443.

D'Amelia, R., and Jacklin, P. 1990. Polyvinyl oleate as a fat replacement. US Pat. 4,915,974.

De Lesser, E. 1993. The making — and remaking — of a 'light' potato chip. Wall Street Journal *91*, No. 43, B6.

Deis, R. 1993. Low-calorie and bulking agents. Food Technology *47*, No. 12, 94.

Dillon, P. 1995. Food nutritional claims. Food Formulating *1*, No. 9, 37.

Dornblaser, L. 1992. Not for the faint-of-heart. Bakery Prodn. and Mktg. *27*, No. 11, 136.

Dougherty, M., Sombke, R., Irvine, J., and Rao, C. 1988. Oat fibers in low calorie breads, soft-type cookies, and pasta. Cereal Foods World *33*, No. 5, 424.

Dreese, P. 1984. Effect of salt level on processing and flavor of white pan bread. Am. Inst. of Baking Tech. Bull. *6*, No. 8, 1.

Dreher, M. 1987. Handbook of Dietary Fiber: An Applied Approach. Marcel Dekker, NYC

Duncan, C. 1990. Method of making a low fat food item having the taste and flavor of a fried food product. U.S. Pat. 4,917,912.

Duxbury, D. 1992. Sunflower oil high in monounsaturates. Food Processing *53*, No. 2, 62.

Duxbury, D. 1993A. Stabilized flax: High-fiber protein. Food Processing *54*, No. 4, 93.

Duxbury, D. 1993B. Uncovering truth about food allergies. Food Processing *54*, No. 2, 52.

Dziezak, J. 1986. Sweeteners in product development. *40*, No. 1, 111.

Dziezak, J. 1989. Fats, oils, and fat substitutes. Food Technology *43*, No. 7, 66.

El-Noklay, M., et al. 1995. Nondigestible fats. US Patent 5,371,254.

Erdman, J. 1990. The quality of microparticulated protein. J. Am. Coll. Nutrition *9*, 398.

Erickson, M., and Frey, N. 1994. Property-enhanced oils in food applications. Food Technology *48*, No. 11, 63.

Fahlen, A. 1990. High fiber natural bread. US Patent 4,971,823.

Fallat, M., et al. Short term study of sucrose polyester, a nonabsorbable fat-like material, as a dietary agent for lowering plasma cholesterol. Am. J. Clinical Nutr. *29*, 1204.

Fan, S. 1991. Ready-to-eat cereal of reduced sodium content and method of preparation. US Pat. 4,988,521.

Farley, D. 1993. Making sure hype does't overwhelm science. FDA Consumer *27*, No. 9, 9.

Fleming, K. 1992. Diet research system. Food Technology *46*, No. 6, 156.

Forand, K. 1991. Cotton fiber particles for use in baked goods. U.S. Pat. 5,026,569.

Franklin, K. 1994. Reduced fat peanut butter and method. US Patent 5,302,409.

Franssell, D., and Palkert, P. 1991. Dry mix for microwave muffins with psyllium and method of preparation. U.S. Pat. No. 5,015,486.

Freeman, T. 1989. Sweetening cakes and cake mixes with alitame. Cereal Foods World *34*, 1013.

Friedman, M. 1995. Way to score in '94. Prepared Foods *164*, No. 5, 44. Patent 4,808,626.

Frübeck, G. 1996. Dietary fiber and coronary heart disease prevention. J. Am. Med. Assoc. *275*, 1883.

Frye, C. 1995. Fat and oil replacers. USP Patent 5,466,473.

Garleb, K., Chmura, J., Anloague, P., Cunningham, M., Sertl, D. 1992. Blend of dietary fiber for nutritional products. U.S. Patent 5,085,883.

Gelardi, R. 1987. The multiple sweetener approach and new sweeteners on the horizon. Food Technology *41*, No. 1, 123.

Giese, J. 1993. Alternative sweeteners and bulking agents. Food Technology *47*, No. 1, 14.

Giese, J. 1995. Vitamin and mineral fortification of foods. Food Technology *49*, No. 5, 110.

Giese, J. 1996. Fats, oils, and fat replacers. Food Technology *50*, No. 4, 78. 1989. Tailored triglycerides having improved autoignition characteristics. U.S. Pat. 4,832,975.

Giese, J. 1996. Olestra: Properties, regulatory concerns, and applications. Food Technology *50*, No. 3, 130.

Gillette, M. 1985. Flavor effects of sodium chloride. Food Technol. *39*, No. 6, 47.

Glass, B., Murphy, M., and Santori, N. 1992. Reduced fat ready-to-spread frosting. U.S. Pat. 5,102,680.

Glicksman, M. 1991. Hydrocolloids the search for oily grail. Food Technol. *45*, No. 10, 94.

Glicksman, M., et al. 1985. Process for a quality reduced-calorie cake. U.S. Pat. 4,526,799.

Goldenberg, R., and Tamura, T. 1996. Prepregnancy weight and pregnancy outcome. J. Am. Med. Assoc. *275*, 1127.

Gori, G. 1988. Weight-control formulation. US Patent 4,784,861.

Gorton, L. 1995. Fiber builders. Baking & Snack *17*, No. 2, 46.

Gregory, J. III. 1988. Recent developments in methods for the assessment of vitamin bioavailability. Food Technology *42*, No 10, 230.

Grimm, R., et al. 1996. Long-term effects on plasma lipids of diet and drugs to treat hypertension. J. Am. Med. Assoc. *275*, 1549.

Grunwald, H., and Rosner, F. 1996. Folic acid fortification of food. JAMA *275*, NO. 9, 662.

Gwynne, J. 1991. Measuring and knowing. The trouble with cholesterol and decision making. J. Am. Med Assoc. *266*, 1696.

Hall, J. 1989. Fat is more than just ugly calories. Analytical Progress *6*, No. 1, 1.

Hammond, N. 1994. Stabilized rice bran. US Pat. 5,376,390.lkkkkk

Hardinge, F., and Hardinge, M. 1993. Vegetarian perspective. Food Tech. *46*, No.10, 114.

Harland, B., and Hecht, A. 1985. Grandma called it roughage. U.S. Dept. Health and Human Services Publ. No. (FDA) 78-2087.

Hasler, C. 1996. Functional foods: Hurdles impeding their development and commercialization. Abstracts Div. of Ag. and Food Chem., Am. Chem. Soc. Annual Meeting *211*,

Havens, A. 1992. System for producing low cholesterol eggs and feed additive resulting from same. US Pat. 5,091,195.

Heckert, D. 1988. Fruit juice beverages and juice concentrates nutritionally supplemented with calcium. U.S. Pat. 4,722,847.

Hefle, S. 1996. The chemistry and biology of food allergens. Food Technology *50*, No. 3, 86.

Hegenbart, S. 1992. In the eye of the storm: Using fiber in food products. Food Product Design 2, No. 1, 19

Hegenbart, S. 1993. Nutraceutical reality on the horizon. Food Prod. Design *3*, No. 9, 19.

Heidinger, K. et al. 1995. Coeliac disease first diagnosed after a bleeding complication. *Reviewed in* J. Am. Med. Assoc. *275*, 894m.

Heidolph, B., and Gard, D. 1995. Leavening composition. US Pat. 5,409,724.

Hess, J. 1976. Snack foods as a nutritional vehicle. Snack Food *65*, No. 9, 15.

Hingley, A. 1993. Food allergies. When eating is risky. FDA Consumer *27*, No. 10, 27.

Hippleheuser, A., Landberg, L., and Turnak, F. 1995. A system approach formulating a low-fat muffin. Food Technology *49*, No. 3, 74.

Hirao, M., Hijiya, H., and Miyaka, T. 1990. Food containing anhyrous crystals of maltitol, and the whole crystalline hydrogenated starch hydrolysate. US Patent 4,917,916.

Holmgren, L. 1988. Preparing a water absorbing dietary fiber product. US Pat. 4,765,994.

Holmgren, L. 1990. Dietary fiber product. US Patent 4,915,960.

Hood, L., and Campbell, L. 1990. Developing reduced calorie bakery products with sucralose. Cereal Foods World *35*, 1171.

Hood, L., and Schoor, M. 1990. Evolution, properties, and applications of an approved high intensity sweetener. Cereal Foods World *35*, 1184.

Hosoda, S., Hosoda, Y., and Kato, E. 1996. Imitation rice and process for making it. US Patent 5,498,435.

Hsu, J., and Larson, G. 1990. Treatment of powdered cellulose. US Patent 4,927,662.

Huang, Y., and Ang, C. 1993. Vegetarian foods for Buddhists. Food Tech. *46*, No. 10, 94.

Hudson, C. Chiu, M., and Knuckles, B. 1992. Development and characteristics of high-fiber muffins with oat bran, rice bran, or barley fiber fractions. Cereal Foods World *37*, 373.

Hussaini, M., and Sakr, A. 1983. Islamic Dietary Laws and Practices. Islamic Food and Nutrition Council of America, Bedford Park, IL.

Hwang, S. 1993. Kosher-food firms dive into mainstream. WSJ *91*, No. 63, B1.

Iijima, Y., et al. 1990. Sweetened condensed milk like composition and a method for producing it. US Patent 4,948,616.

Ingeno, C. 1996. The ginseng zing. Food Product Design *6*, No. 3, 24.

Ingersoll, B. 1990. FDA demands firms support oat bran. Wall Street J. *86*, 42, B5.

Inglett, G. 1991. Method for making a soluble dietary fiber composition from oats. U.S. Pat. 4,996,063.

Inglett, G. 1996. Hydrolyzed oat flour used in low fat foods. Abstracts Am. Chem. Soc. Div. Ag. and Food Chem. Annual Mtg. *211*.

Inglett, G., and Grisamore, S. 1991. Maltodextrin fat substitute lowers cholesterol. Food Technology *45*, No. 6, 104.

Irwig, L., Glaziou, P., Wilson, A., and Macaskill, P. 1991. Estimating an individual's true cholesterol level and response to intervention. J. Am. Medical Assoc. *266*, 1678.

Irwin, W. 1990. Isomalt — a sweet bulking agent. Food Technology *44*, No. 6, 128

Iyengar, R., Zaks, A., and Gross, A. 1991. Starch-derived, food-grade, insoluble bulking agent. US Patent 5,051,271.

Izzo, M., Stahl, C., and Tuazon, M. 1995. Using cellulose gel and carrageenan to lower fat and calories in confections. Food Technology *49*, No. 7, 45.

Jackel, S. 1987. Calcium fortification. Proc. Am. Soc. Bakery Engrs. *1987*, 53.

Johnson, H. 1977. Low cholesterol diet: it won't prevent heart attacks. Dun's Review *March*, 93.

Johnson, L. 1994. Vitamin and mineral fortification of foods. Food Tech. *48*, No. 7, 124.

Jones, J. 1990. Oat bran in the news. Cereal Foods World *35*, 515.

Jones, J. 1991A. Cancer and diet — controversy continues. Cereal Foods World *36*, 832.

Jones, J. 1991B. Medium-chain triglycerides. Cereal Foods World *36*, 793.

Jongh, G., et al. 1968. Bread without gluten. Bakers Dig. *1968*, No. 6, 24.

Kadan, R., and Ziegler, G., Jr. 1986. Effects of ingredients on iron solubility and chemical state in experimental breads. Cereal Chem. *63*, 47.

Kasarda, D. 1978. The relationship of wheat proteins to celiac disease. Cereal Foods World *23*, 240.

Kaslas, P., et al. 1994. Fried food with less oil. US Patent 5,312,635.

Katz, F. 1996A. The medical mix. Food Processing *57*, No. 6, 75.

Katz, F. 1996B. New processing techniques test the water. Food Processing *53*, No. 2, 66.

Katz, S. 1996. Are essential micronutrients toxic? Food Testing & Analysis *2*, No. 3, 19.

Kehoe, D., Kong, L., Tisdel, P., MacKenzie, K., Hall, R., Field, and Rao, C. 1990. Subchronic safety study of bleached oat hull fiber in rats. Cereal Foods World *85*, 1026

Kevin, K. 1995A Naturally sweet. Food Processing *56*, No. 10, 93.

Kevin, K. 1995B. Phascinating phytochemicals. Food Processing *56*, No. 4, 79.

Kevin, K. 1995C. Starch de resistance. Food Processing *56*, No 1, 65.

Kevin, K. 1996. Whey-derived sweetener goes commercial. Food Processing *57*, No. 2, 88.

Kilara, A., and Iya, K. 1993. Food and dietary habits of Hindu. Food Tech. *46*, No. 10, 94.

Kinsella, J. 1988. Food lipids and fatty acids: Importance in food quality, nutrition, and health. Food Technology *42*, No. 10, 124

Klemann, L., and Finley, J. 1989. Low calorie fat mimetics comprising carboxy/carboxylate esters. US Patent 4,830,787.

Klemann, L., Finley, J., and Scimone, A. 1990. Low calorie fat mimetics. U.S. Pat. 4,959,465.

Klemann, L., et al. 1991A. Cyclohexyl diol diesters as fat mimetics. U.S. Pat. 5,006,351.

Klemann, L., et al. 1991B. Long chain diol diesters as mimetics. U.S. Pat. 5,008,126.

Klemann, L., et al. 1991C. Thioester derivatives as fat mimetics. U.S. Pat. 4,992,293.

Klis, J. 1996. Oat bran is back. Food Technology *50*, No. 3, 180.

Koch, R. 1993. Fiber. Proc. Am. Soc. Bakery Engrs. *1993*, 73.

Kolata, G. 1983. Dietary dogma disproved. Nutritionists find that some complex carbohydrates act like simple sugars and vice versa. Science *220*, 487.

Kosmark, R. 1996. Salatrim: Properties and applications. Food Technology *50*, No. 4, 98.

Krogull, M. 1992. Dietary fiber . . . increased consumption and interest. Food Science Newsletter (Hazelton Labs) No. 44.

Kroskey, C. 1995. Cakes and pies:Death by chocolate. Bakery Prodn. Mktg. *30*, No. 1, 32.

Kuhn, M. 1994. A season for cereal. Food Business *7*, No. 4, 60.

Kuhn, M. 1996A. Betting on fat-free. Food Processing *53*, No. 2, 64.

Kuhn, M. 1996B. Soy in the spotlight. Food Processing *57*, No. 5, 52.

Kuntz, L. 1996A. Ein prosit! (To your health). Food Product Design *6*, No. 1, 69.

Kuntz, L. 1996B. Designing nature's way. Food Product Design *5*, No. 11, 79

Kurtzweil, P. 1994. Making it easier to shed pounds. FDA Consumer *28*, No. 6, 10.

Kwapong, O., and Fedungjacklin, V. 1996. Hunger suppressant beverage. US Patent 5,472,716.

LaBell, F. 1993. Experts ask policy change on health claims. Food Proc. *54*, No. 4, 52.

LaBell, F. 1995A. Flavor and stabilizer duos. Food Development *164*, No. 9, 15.

LaBell, F. 1995A. Ensuring kosher quality. Prepared Foods *164*, No. 7, 66.

LaBell, F. 1995B. Functional nutritional fiber. Prepared Foods *164*, No. 11, 87.

LaBell, F. 1996A. Low-calorie tuber flour for pasta, baked goods. Food Proc. *53*, No. 4, 56.

LaBell, F. 1996B. Taking the bite out of lowfat dressings. Prepared Foods *165*, No. 6, 109.

Labin-Goldscher, R., and Edelstein, S. 1996. Calcium citrate: A revised look at calcium fortification. Food Technolgoy *50*, No. 6, 96.

LaChance, P. (Editor) 1994. Human obesity. Food Technology *48*, No. 2, 127.

Larkin, M. 1994. Lowering cholesterol. FDA Consumer *28*, No. 2, 26.

Larkin, M. 1996. Health information on-line. FDA Consumer (June 1996) 23.

Lasdon, L., et al. 1989. Low calorie peanut spread. U.S. Pat. 4,828,868.

Lecos, C. 1983. A compendium on fats. HHS Publication No. (FDA) 83-2171. Public Health Service, Rockville, MD

Lecos, C. 1985. Diet and the elderly. HHS Publication No. (FDA) 85-2201. Public Health Service, Rockville, MD.

Lee, E., and Tandy, I. 1996. Salt-taste enhancement. US Patent 5,494,689.

Lee, K., and Clydesdale, F. 1980. Effect of baking on the forms of iron in iron-enriched flour. J. Food Sci. *45*, 1500.

Lehrer, S., and O'Neil, C. 1992. Occupational reactions in food. Food Tech. *46*, No. 5, 153. Technology *42*, No. 10, 194.

Lenfant, C. 1996. High blood pressure. Some answers, new questions, continuing challenges. J. Am. Med. Assoc. *275*, 1604.

Levin, C. 1991. Medium-chain triglycerides. Cereal Foods World *36*, 793.

Lewis, V., and Lewis, D. 1995. Preparing no- or low-fat potato chips and straws. US Pat. 5,441,758

Lillford, P. 1994. Food research trends. World of Food Ingredients. *1994*, (Oct-Nov.), 32.

Lingle, R. 1987. New Tech's high-tech, low-calorie snacks. Prepared Foods *156*, No. 12, 76.

Liu, Y. 1996. Immune function of curcumin. Abstracts 211th Annual Mtg. Am. Chem. Soc. Div. Ag. and Food Chem. *1996*

Low, L., Mackerer, C., Feuston, M., and Komineni, C. 1994. Synthetic cooking oil. US Patent 5,320,857.

Lupton, j., and Yung, K. 1991. Interactive effects of oat bran and wheat bran on serum and liver lipids and colonic physiology. Cereal Foods World *36*, 827,

Madsen, K. 1981. The anticaries potential of seeds. Cereal Foods World *26*, No. 1, 19.

Marlett, J. 1991. Dietary fiber content and effect of processing on two barley varieties. Cereal Foods World *36*, 576.

Mattson, J., et al. 1976. The effect of a nonabsorbable lipid, sucrose polyester, on the absorption of dietary cholesterol by the rat. J. Nutrition *29*, No. 6.

McCarron, D., and Hatton, D. 1996. Dietary calcium and lower blood pressure. We can all benefit. J. Am. Med. Assoc. *275*, 1129.

McCoy, S., Madison, B., Self, P., and Weisgerber, D. 1989. Sucrose polyesters which behave like cocoa butters. U.S. Pat. 4,822,875.

McCue, N. 1996. Meat flavors for vegetarian fare. Prepared Foods *165*, No. 6, 111.

McGinley, L. 1996. US orders makers of grain products to add folic. WSJ *97*, No. 43, B-12.

McGinley, L., and Narisetti, R. 1996. Fat substitute of P&G wins FDA approval. Wall Street J. *97*, No. 18, A-3.

McNaughton, K. 1988. Cholesterol is not a proven killer. Industrial Chemist *9*, No. 1, 10.

Megremis, C. 1991. Medium-chain triglycerides: A nonconventional fat. Food Technology *46*, No. 2, 108.

Meibach, R. 1996. Cholesterol-reduced egg products. U.S. Pat. 5,487,912.

Metcalfe, D. D. 1992. The nature and mechanisms of food allergies and related diseases. Food Technol. *46*, No. 5, 135.

Meyer, A. 1990. Heartburn for the AHA. Prepared Foods *159*, No. 1, 15.

Midgley, J., Matthew, A., Greenwood, C., and Logan, A. 1996. Effect of reduced dietary sodium on blood pressure. J. Am. Med. Assoc. *275*, 1590.

Mijac, M., and Guffey, T. 1989. Baked goods made with sucrose fatty acid esters. U.S. Pat. 4,835,001.

Miller, R. 1981. On being too rich, too thin, too cholesterol laden. HHS Publication (FDA) 81-1087.

Miller, W. 1987. The legacy of cyclamate. Food Technology *41*, No. 1, 116.

Mongeau, R., and Brassard, R. 1990. Determination of insoluble, soluble, and total dietary fiber: Collaborative study of a rapid gravimetric method. Cereal Foods World *35*, 319.

Moore, E., and McAnalley, B. 1996. Aloe vera drink. US Patent 5,443,830.

Morck, T., and Cook, J. 1981. Factors affecting the bioavailability of dietary iron. Cereal Foods World *26*, 667.

Morrison, H., et al. 1996. Serum folate and risk of fatal coronary heart disease. J. Am. Med. Assoc. *275*, 1893.

Moskowitz, A. 1988. Highly palatable dietary fiber supplement. U.S. Pat. 4,766,004.

Mossoba, M., and Firestone, D. 1996. New methods for fat analysis in foods. Food Testing & Analysis *2*, No. 2, 24.

Muir, J., Birkett, A., Phillips, J., Jones, G., and O'Dea, K. 1996. Resistant starch — implications for colonic health. Am. Assoc. Cereal Chem. Annual Mtg. Abstract *1996*, 164.

Navicki, L., and Nielsen, T. 1992. Mixing and extensional properties of wheat flour doughs with added corn flour, fibers, and gluten. Cereal Foods World *37*, No. 1, 37.

Nelson, G. (Editor) 1991. Health Effects of Dietary Fatty Acids. Am. Oil Chemists Soc., Champaign, IL

Nelson, K. 1990. Fat substitutes in baking. Proc. Am. Soc. Bakery Engrs. *1990*, 79.

Newman, B. 1993. Kosher vodka is winning converts in Poland. WSJ *91*, No. 109, A1.

Newsome, R. 1986. Sweeteners: Nutritive and non-nutritive. Food Techn. *40*, No. 8, 195.

Newton, S. 1989. Fats and Oils: How to they perform? Prepared Foods *158*, No. 5, 178.

Nicolosi, R. 1991. Rice bran oil research unveiled. Rice Bran News *3*, No. 1, 1.

Ning, L., Villota, R., and Artz, W. 1991. Modification of corn fiber through chemical treatments in combination with twin-screw extrusion. Cereal Chem. *68*, 632,

O'Donnell, C. 1991. Can technology temper snack wars? Prepared Foods *160*, No. 12, 55.

O'Donnell, C. 1993. Kosher confusion. Prepared Foods *162*, No. 3, 55.

O'Donnell, C. 1994A. Global ingredient trends and advances. Prep. Foods *163*, No. 1, 46.

O'Donnell, C. 1994B. Meeting consumer demands. Prepd. Foods *163*, No. 6, 42.

O'Donnell, C. 1995. Great oats! Fiber's comeback. Food Prod. Development *165*, No. 3, 65.

O'Donnell, C. 1996. The changing world of fats and oils. Prepared Foods *165*, No. 7, 36.

Katz, F. 1996. Special delivery. Food Processing *57*, No. 6, 66.

Obarzanek, E., Velletri, P., and Cutler, J. 1996. Dietary protein and blood pressure. J. Am. Med. Assoc. *275*, 1598.

Ohren, J. 1973. Protein fortification with cottonseed flour. Snack Food October 1973, p. 38

Oku, 1994. Metabolism of sugar alcohols. Presented 28 June 1994, Carbohydrate Division Symposium, Institute of Food Technologist Convention.

Ono, Y. 1994. Let there be 'right.' Food marketers seek appetizing alternatives to 'dark.' Wall Street Journal *94*, No. 24, B1.

Ono, Y., and Quick, R. 1995. Modern-day alchemy: the low-everything cookie. Wall Street J. *96*, No. 98, B1.

Ory, R. 1991. Grandma Called It Roughage. American Chemical Society, Washington.

Otto, A. 1993. Putting taste into healthy soy-based foods. Food Business 6, No. 13, 50.

Pacyniak, B. 1992. Born again kosher. Prepared Foods 167, No. 9, 15.

Pape, S. 1996A. Changes in labeling solve allergen dilemma? Prepd. Fds. 165, No. 4, 32.

Pape, S. 1996B. Olestra exposes flaws of FDA review. Prepared Foods 165, No. 3, 26.

Park, H., Seib, P., and Chung, O. 1995. Effects of wheat fiber on breadmaking performance. Am. Assoc. Cereal Chemists Convention 1995.

Patton, J., et al. 1986. Reduced calorie yeast leavened product. US Patent 4,587,126.

Pflaumer, P., et al. 1990. Cookies containing psyllium. US Patent 4,950,140.

Pitts, E. 1994. Dairy ingredients. New opportunities in functional foods. The World of Food Ingredients 1994 (Oct-Nov) 40.

Pomeranz, Y., and Meloan, C. 1978. Food Analysis: Theory and Practice. Van Nostrand Reinhold, NYC.

Ponte, J., Jr., and Walker, C. 1995. Sensory evaluation of high fiber pie crust. Am. Assoc. Cereal Chemists Convention 1995.

Potter, N. 1986. Food Science. AVI Publishing Co., Westport, CT

Prosky, I., et al. 1984. Determination of total dietary fiber in foods and food products. J. Assoc. Official Anal. Chem. 67, 1044.

Pszczola, D. (Editor) 1992A. The nutraceutical initiative: A proposal for economic and regulatory reform. Food Technology 46, No. 4, 77.

Pszczola, D. 1992B. The nutraceutical initiative: A proposal for regulatory reform. Food Technology 46, No. 4, 77.

Pszczola, D. 1994. Packaging systems reflect refinement. Food Technol. 48, No. 8, 99.

Raber, L. 1995. FDA panel probes safety of fat substitute. C&EN 73, No. 47, 11.

Ramaswamy, S. 1991. Fiber and method of making. U.S. Pat. 5,023,103.

Ramsay, M., et al. 1991. Processing rice bran by supercritical fluid extraction. Food Technology 45, No. 11, 98.

Ranhotra, G., Gelroth, J., and Bright, P. 1988. Effect of the sources of fiber in bread on intestinal responses and nutrient digestibilities. Cereal Chem. 65, 9.

Ranhotra, G., et al. 1990. Relative lipidemic responses in rats fed oat bran or oat bran concentrate. Cereal Chem. 67, 509.

Ranhotra, G., Gelroth, J., Glaser, B., and Posner, E. 1992. Total and soluble fiber content of air-classified white flour from hard and soft wheats. Cereal Chem. 69, 75.

Rapaille, A., Gonze, M., and Van der Shureren, F. 1995. Formulating sugar-free chocolate products with maltitol. Food Technology 49, No. 7, 51.

Rawls, R. 1981. New research muddies cholesterol debate. C&EN 26 Jan 81, 37.

Recker, R. 1988. Calcium absorbability from milk products, an imitation milk, and calcium carbonate. Am. J. Clinical Nutr. 47, No. 1, 93.

Regelson, W. 1994. The Melatonin Miracle. Simon & Schuster, NYC

Reynolds, P. 1996. Ener-G goes glass. Packaging World 3, No. 3, 51.

Rimm, E., Ascherio, A., and Willett, W. 1996. Dietary fiber and coronary heart disease prevention. J. Am. Med. Assoc. 275, 1883.

Rimm, E., et al. 1996. Vegetable, fruit, and cereal fiber intake and risk of coronary heart disease among men. J. Am. Med. Assoc. 275, 447.

Rinse, J. 1978. Cholesterol and phospholipids in relation to atherosclerosis. Am. Laboratory 10, No. 4, 68.

Roberfroid, M. 1995. A functional food. The World of Ingredients 1995, March-April, 42.

Rodricks, J. 1996. Safety assessment of new food ingredients. Food Tech. 50, No. 3, 114.

Sacks, F. 1991. The role of cereals, fats, and fibers in preventing coronary heart disease. Cereal Foods World 36, 822.

Saikusa, T. 1996. Gamma-butyric acid-enriched food. U.S. Pat. 5,472,730.

Sakai, H., and Kawakita, T. 1991. Proteins: Amino acids. *In* Encyclopedia of Food Science and Technology. John Wiley & Sons, NYC

Sampson, H. 1992. Food hypersensitivity: Manifestations, diagnosis, and natural history. Food Technol. *46*, No. 5, 141.

Sanders, S. 1993. Dried plums: a multi-functional bakery ingredient. Am. Soc. Bakery Engineers Bulletin No. *228*,

Sanderson, G. 1996. Gums and their use in food systems. Food Technology *50*, No. 3, 81.

Sbczynska, D., and Setser, C. 1991. Replacement of shortening by maltodextrin-emulsifier combinations in chocolate layer cakes. Cereal Foods World *36*, 1017.

Schneeman, B. 1986. Dietary fiber: physical and chemical properties, methods of analysis, and physiologival effects. Food Technology *40*, No. 2, 104.

Schneeman, B. 1989. Dietary fiber. Food Technology *43*, No. 10, 133.

Schroder, B., and Baer, R. 1991. Consumer evaluation of reduced-cholesterol butter. Food Technology *45*, No. 10, 104.

Seligson, F., Hunter, J., and St. Clair, A. 1988. Food composition with superior blood cholesterol lowering properties. US Pat. 4,789,664.

Seyam, A. 1996. Fat-free or low-fat cookie production. US Pat. 5,492,710

Shaw, G., Velie, E., and Schaffer, D. 1996. Risk of neural tube defect — affected pregnancies among obese women. J. Am. Med. Assoc. *275*, 1093.

Shaw, J., and Sharma, S. 1989. Ingestible aggregate and delivery system prepared therefrom. U.S. Patent 4,851,392.

Shinnick, F., Mathews, R., and Ink, S. 1991. Serum cholesterol reduction by oats and other fibers. Cereal Foods World *36*, 815.

Siguel, E., and Lerman, R. 1996. The effect of low-fat diet on lipid levels. J. Am. Med. Assoc. *275*, 759.

Silge, M. 1992. Dried-fruit-based fat replacement systems.

Silva, R. 1991. High fiber bakery products. U.S. Pat. 5,035,903.

Singer, N., Latella, J., and Shoji, Y. 1990. Fat emulating protein products and process. U.S. Pat. 4,961,953.

Siscovick, D., et al. 1996. Dietary sources of long-chain n-3 polyunsaturated fatty acids. J. Am. Med. Assoc. *275*, 837.

Skerritt, J., and Hill, A. 1992. How free is gluten-free? Relationship between Kjeldahl nitrogen values and gluten protein content for wheat starches. Cereal Chem. *69*, 110.

Skerritt, J., Devery, J., and Hill, A. 1990, Gluten intolerance: Chemistry, celiac-toxicity, and detection of prolamins in foods. Cereal Foods World *35*, 638.

Smith, T. 1992. All the food you can eat. Today's Chemist at Work 2, No. 3, 16.

Smith, T. 1996. Do sports drinks work? Today's Chemist at Work 5, No. 6, 36.

Sosulski, F., and Wu, K. 1988. High-fiber breads containing field pea hulls, wheat, corn, and wild oats brans. Cereal Chem *55*, 186.

Spizziri, J. 1995. Formulating for special needs. Food Product Design 5, No. 7, 29.

Stampfer, M. 1996. Folate and cardiovascular disease. J. Am. Med. Assoc. *275*, 1929.

Steginck, L. 1987. Aspartame: Review of the safety issues. Food Technology *41*, No. 1, 119.

Stipp, D. 1990. Oat bran's cholesterol benefits are challenged. Wall St. J. 18 Jan 90, B-1.

Sullivan, D. 1994. Fatty acids take prominent role in new regulations, research. Hazleton Laboratories Food Science Newsletter *51*.

Taylor, S. 1992. Chemistry and detection of food allergens. Food Technol. *46*, No. 5, 146.

Taylor, S., and Nordlee, J. 1996. Detection of food allergens. Food Tech. *50*, No. 5, 231.

ter Haseborg, E., and Himmelstein, A. 1988. Quality problems with high fiber breads solved by use of hemicellulose enzymes. Cereal Foods World *33*, 419.

Thayer, A. 1992. Food additives. Chem. & Eng. News *70*, No. 24, 26.

Toops, D. 1996A. Herbal odysseys. Food Processing *57*, No. 6, 46.

Toops, D. 1996B. Xyrofin plants roots in the USA. Food Processing *57*, No. 1, 37.

Troncone, R., and Auricchio, S. 1991. Gluten-sensitive enteropathy (celiac disease). Food Reviews International 7, 205.

Trowell, H. 1972. Crude fiber, dietary fiber and atherosclerosis. Atherosclerosis *16*, 138.

Tyrpin, H., et al. 1996. Erythritol sweetened chewing gum. US Patent 5,494,685.

Ueda, N, and Takaichi, A. 1991. High protein nutritive food and process for preparing same. U.S. Pat. 5,051,270.

Vaisey-Genser, M. 1994. Flaxseed. Health, nutrition, and functionality. The Flax Council of Canada, Winnipeg.

Venardos, J. 1996. Private communication. Cultor Food Science, NYC

Verduin, P. 1991. Process for making low sodium sponge goods and the products obtained thereby. US Pat. 5,064,661.

Villagran, M., et al. 1995. Process for making reduced-fat fried snacks. US Pat. 5,464.642.

Vollendorf, N., and Marlett, J. 1991. Dietary fiber methodology and composition of oat groats, bran, and hulls. Cereal Foods World *36*, 565.

Volpenhein, R. 1986. Acylated glycerides useful in low calorie fat-containig food compositions. US Pat. 4,582,715.

Wagner, J. 1992A. Global consumers want the lite stuff. Food Processing *53*, No. 10, 68.

Wagner, J. 1992B. Lite 'n healthy goes mainstream. Food Business *5*, No. 21, 10.

Wagner, J. 1994. Ten replacements for the word "light" on labels. Food Proc. *55*, No. 9, 68,

Wahn, U., Wahn, R., and Rugo, E. 1992. Comparison of the residual allergenic activity of six different hydrolyzed protein formulas. J. Pediatrics *121*, S80.

Wall, J., and Carpenter, K. 1988. Variation in availability of niacin in grain products. Food Technology *42*, No. 10, 198.

Washen, C. 1988. Increasing dietary fiber to treat diseases. Cereal Foods World *33*, 452.

Waslien, C. 1988A. Salt restriction for healthy children. Cereal Foods World *33*, 452.

Waslien, C. 1988B. What is dietary fiber and what is starch? Cereal Foods World *33*, 312.

Weaver, C. 1988. Calcium and hypertension. Cereal Foods World *33*, 792.

Weiss, T. 1970. Food Oils and Their Use. AVI Publ. Co., Westport, CT

Werler, M., Louik, C., Shapiro, S., and Mitchell, A. 1996. Prepregant weight in relation to risk of neural tube defects. J. Am. Med. Assoc. *275*, 1089.

Williams, R. 1994. FDA proposes folic acid formulations. FDA Consumer *28*, No. 4, 11.

Winslow, R. 1996. Culprit in coronary artery disease might be a bacterium, study suggests. Wall Street J. *97*, No. 108, B8.

Woodbury, M., and Woodbury, M. 1993. Neuropsychiatric development: two case reports about the use of dietary fish oils and/or choline supplementation in children. J. Am. Coll. Nutrition *11*, 473.

Worthy, W. 1990A. Evidence mounts for dietary soluble fiber benefits. Chem. Eng. News *68*, No. 22, 23.

Worthy, W. 1990B. New sweet, salty peptides synthesized. C&EN *68*, No. 2, 25.

Worthy, W. 1991. Low-calorie fat replacer is based on pectin. C&EN *69*, No. 42, 26.

Yackel, W., and Cox, C. 1992. Application of starch-based fat replacers. Food Technology *46*, No. 6, 146.

Yang, H., et al. 1996. Sterol esterification in yeast: A two-gene process. Science *272*, 1353.

Ye, Q., Heck, G., and DeSimone, J. 1991. The anion paradox in sodium taste reception: Resolution by voltage-clamp studies. Science *254*, 724.

Young, J., et al. 1992. Reduced calorie potato chips and other low moisture fat-containing foods having less waxiness and improved flavor display. US Pat. 5,085,884.

Yuon, E., and Chien, F. 1996. Asian American and Pacific Islander health: A paradigm for minority health. J. Am. Med. Assoc. *275*, 735.

Zook, D. 1992. Process for infusion of partially defatted nuts. U.S. Pat. 5,094,874.

INDEX

www.ingramcontent.com/pod-product-compliance
Lightning Source LLC
Chambersburg PA
CBHW021030210326
41598CB00016B/966